THE
Upright
Ape

A NEW ORIGIN
of the SPECIES

Dr. Aaron G. Filler

NEW PAGE BOOKS
A division of The Career Press, Inc.
Franklin Lakes, NJ

THE UPRIGHT APE
EDITED BY KIRSTEN DALLEY
TYPESET BY EILEEN DOW MUNSON
Cover design by Howard Grossman/12e
Printed in the U.S.A. by Book-mart Press

To order this title, please call toll-free 1-800-CAREER-1 (NJ and Canada: 201-848-0310) to order using VISA or MasterCard, or for further information on books from Career Press.

The Career Press, Inc., 3 Tice Road, PO Box 687,
Franklin Lakes, NJ 07417
www.careerpress.com
www.newpagebooks.com

Library of Congress Cataloging-in-Publication Data

Filler, Aaron G., 1956-
    The upright ape : a new origin of the species / by Aaron G. Filler.
        p. cm.
    Includes bibliographical references and index.
    ISBN-13: 978-156414-933-6
    ISBN-10: 1-56414-933-1
        1. Evolution (Biology) 2. Bipedalism. 3. Apes—Locomotion. I. Title.

QH366.2.F545 2007
576.8—dc22

                                                        2007011942

*This book is dedicated to the memory of
Ernst Mayr (1904–2005)
and
Stephen Jay Gould (1941–2002),
both of whose contributions to science were as
legion as they were magnificent.*

Acknowledgments

I have been blessed with the greatest teachers, including Stephen Jay Gould, Ernst Mayr, Irven DeVore, David Pilbeam, Farish Jenkins, A.W. Crompton, Len Radinsky, Leigh VanValen, and Russell Tuttle, among many others. In addition, my research was conducted at fantastic institutions—UCLA, Harvard, Cambridge, and the University of Chicago. I continue to be inspired by brilliant colleagues and friends, in particular Terry Deacon at Berkeley, Andrew Lever at Cambridge, and Pat Johnson at Cedars Sinai.

Although the idea of this book has grown steadily over more than 30 years of research, I credit the excitement of Joan Bossert, Vice President of Oxford University Press, for convincing me that it was time to start writing it. However, it is the tireless efforts and enthusiasm of my agent, Jodie Rhodes, and the intrepid spirit of Michael Pye and his team at New Page Books—most notably my editor, Kirsten Dalley—that have made a book of this scope possible.

To write a work such as this while maintaining an intense and fast-paced world-class neurosurgery practice has been possible because of my remarkably accomplished team led by Sheila Butler, Jodean Petersen, Brie Dean at Cedars Sinai, and our gifted manager, Shirlee Jackson. As a result, we have the best medical technology in the world and can reliably deliver excellent results for our patients from around the world.

My parents inspired me to think expansively and innovate constantly. My wife and kids—Lise, Rachel, and Wyatt—make life a constant joy. All of their patience and support while this project went forward are greatly appreciated.

Contents

Foreword

I first met Aaron Filler when I joined the Harvard faculty in 1981, and enjoyed interacting with this stimulating polymath as he wrestled with his doctoral research topic on the evolution of the variety of ways apes, humans, and their ancestors moved—through the trees and on the ground. How animals move (or "locomote," in technical jargon) is critical to understanding a species' biology, for moving is vital to feeding, finding mates, and avoiding predators, and hence for survival and reproduction. Aaron is now Dr. Filler and a distinguished neurosurgeon. But in the two decades since completing his doctorate, he has continued to think not only about what happened in the locomotor evolution of apes and humans (technically, "hominoids"), but how it happened developmentally, and why it happened in terms of selective advantage.

After a century and a half of controversy, the evolutionary relationships of the few surviving hominoid species are well established through abundant genetic analyses of these living species. Humans are the closest relatives of the two chimpanzee species, whereas the African gorillas (which resemble chimps closely and in many ways resemble enlarged versions of their smaller African cousins) are more distantly related. The large Asian orangutan is still further removed, and the small and acrobatic Asian gibbons are the most distantly related of all the hominoids. It is now believed that these living hominoids share a common ancestor that lived more than 20 million years ago. Although these relationships are now very widely (though not quite universally) accepted, almost every possible combination of relationships among the species had been proposed over the past century.

There have been almost as many suggestions about the locomotor evolution of these species, but almost all have agreed that the defining feature of humans and their ancestors after the split from the chimpanzee lineage was bipedalism: habitual upright walking. To be an habitual biped requires a

7

whole suite of anatomical adaptations (special features shaped by natural selection). The large apes—chimps, gorillas, and orangutans—are occasionally bipedal, but also use many other movement patterns. Among the hominoids, it is the small gibbons that are most frequent bipeds, although other ways of moving are more frequent in their repertoire. More than a century ago, before our modern understanding of hominoid evolutionary relationships, the distinguished anatomist Arthur Keith proposed that bipedalism was a constant theme throughout hominoid evolution, and not just in the evolution of our own lineage. This model has fallen out of favor, and it is now more widely accepted, though not universally, that the common ancestor of humans and chimps would have been quite chimp-like (and gorilla-like), moving in the trees by vertical clambering and arm suspension, and on the ground quadrupedally, with feet flat and toes extended and fingers curled so that the knuckles contact the ground; bipedalism would have been very infrequent.

Dr. Filler brings a refreshingly contrarian approach to these questions. His research involves careful anatomical observation, the analysis of critical fossils, along with developmental biology of the spine—indeed, he was one of the very first to bring the then barely nascent field of evolutionary developmental biology to bear on a problem in human evolution—blending them into a novel theory. Briefly, he argues that bipedalism has always been an important component of the locomotor behavior of all living hominoids since their common ancestor evolved such behaviors (along with corollary and necessary anatomical and developmental adaptations). Dr. Filler discusses in detail the anatomical and developmental bases for his argument. As he points out provocatively, if some significant degree of bipedalism precedes the divergence of humans and chimps, this raises most interesting questions about the recognition, interpretation, and definition of the earliest possible human ancestors.

This book is bound to stimulate; its arguments are challenging, may not be widely accepted, but will need to be taken seriously.

David Pilbeam
Henry Ford II Professor of Human Evolution
Peabody Museum
Harvard University

Preface

Extraordinary problems in science call for extraordinary solutions.

This book tells the new story of evolution—an ever-broadening area of research and theory that includes not only Darwinian ideas, but also the biology of biochemical origins and the genetics of biological innovation, both of which go beyond what has traditionally been considered in classical Evolutionary Theory. In part, this book details what I have discovered since the day in 1981 when David Pilbeam placed in my hands the problem of explaining a seemingly inexplicable 21-million-year-old fossil. This bone had the totally unique features found in humans, but it was from a creature that lived 15 million years too soon. There should not have been anything that looked like this until the human-chimp split 6 million years ago. I believe this conundrum can be explained, but Darwinian Evolutionary Theory as it now stands cannot provide everything that is required to explain it. A broader, more general theory of evolution is now necessary—one that truly incorporates the explosive growth of our scientific knowledge over the past 25 years.

Wearing a sealed "hazmat" suit and sifting through buckets of human sera infected with the HIV virus in 1992, I ran some experiments that may have cracked open the mystery of the link between rapid evolution and environmental change through the enzymatic mechanics of transcriptase enzymes. The lab was in a sealed, air-locked, trailer-like unit on the roof of St. George's Hospital Medical School in London. A few floors below in the MRI lab, I ran experiments in the superpowered 4.7 Tesla experimental MRI system. These experiments led me to discover the first method to image nerves inside the human body—Magnetic Resonance Neurography—a method that I then applied to the problem of understanding evolutionary changes in the number of body segments in humans. The answers revealed by these two threads of research turned out to have little to do with Darwinian evolutionary mechanisms.

I also scoured through thousands of museum specimens in the vast warehouse collection rooms of the Field Museum in Chicago, the American Museum of Natural History in New York, the Smithsonian Museum in Washington, D.C., the Museum of Comparative Zoology at Harvard, and the British Museum in London. I have operated on the spines of thousands of patients to fully understand and repair the ravages of our biological limitations. I have even developed new types of positron emitting labels (matter-antimatter materials) that travel through the nerves by a process called axonal transport, in an effort to unravel the problem of altered similarity among species. The result of all of this has included numerous academic papers, nearly a score of patents, hundreds of invited lectures at major medical and scientific societies, several books, and a wall full of diplomas, awards, and degrees. It's all there on Google, Delphion, PubMed, and Amazon if you want to see the details.

Now, however, the time has come to pull together all of these threads into a single, synthesized work for the scientifically interested public. This book details a new and modernized theory of evolution, one that will revolutionize the very meaning of the word human. I have chosen to make my case by proceeding simultaneously in both the formal academic arena and in the broader sphere of access provided by a book written for the general public. It's important that the historical and philosophical underpinnings of our scientific endeavors as well as the more technical aspects of our experimental tests be widely understood.

Every living human has a stake in knowing who we are and how we came to be. For all these reasons, *The Upright Ape* is as broad and extensive as it is relentlessly controversial. When existing theories can no longer accommodate the evidence at hand, it is time for innovation and regeneration. It is my belief that this book will help lead the way to a new and stronger understanding of nature and of ourselves.

The History of Life

Introduction

The birth of Darwinian Evolutionary Theory was accompanied by the final burial of the competing ideas of the poet-naturalist Johann Wolfgang von Goethe and the great French zoologist Étienne Geoffroy Saint-Hilaire. Darwin himself draws attention to this in his preface to later editions of *The Origin of Species*. Darwin provided a powerful insight into how new species arise and how they become different from their ancestors, whereas Goethe and Geoffroy were concerned with explaining why the disparate species of life retain so many similarities to each other. If the modern molecular biology of the past 20 years definitively proves that Goethe and Geoffroy were correct all along, how can Darwinian Theory stand unchanged?

There is a new and emerging transformation occurring in evolutionary theory today—namely, the inclusion of Modularity Theory (Schlosser and Wagner 2004). This is based in part on insights from our new molecular genetic understanding of how organisms progress through their embryological development. In the past, ideas and suggestions in this area were easily dismissed with rhetorical ripostes. Now, however, the new existence of powerful concrete molecular data has forced Darwinian biologists to come to grips with the conflicting view of Goethe and Geoffroy.

Explosions and Revolutions in the History of Life

This book is concerned with improving our understanding of three "explosions" and three "revolutions" in nature that are responsible for the emergence of most of the organisms that fascinate us in the biological world. From a number of points of view, these six events are the great moments in the History of Life on this planet, and each demonstrates the power of Modularity Theory for clarifying the mechanisms of evolutionary change.

Many biologists—such as the "cladists" (who have applied rigorous logical principles to the science of relationships among organisms)—believe that the History of Life is composed only of millions of individual and equally important branching events among individual species, with no one event more significant than another. Strict adherents to the Modern Evolutionary Synthesis—the "Darwinians"—are comfortable with some significance being attached to these six events, but see no special obligation upon Evolutionary Theory to explain them. A third major group of biologists—the "evodevo" enthusiasts (who look for explanations of evolution in the embryological development of organisms)—are substantially focused on the explanation of only two of these six events.

The Origin of Descent

The first revolution we will look at is the origin of biological descent. In the ancient seas of a billion years ago, transcription and translation to copy DNA and make proteins was already taking place in much the same way it occurs in our own bodies today. However, genes were passed horizontally—that is, from one adult organism directly to other adult organisms of various kinds. At some point in time, a new type of organism appeared that carefully guarded its own set of genes and passed them on only to its progeny vertically, as an intact set with only minimal modification. This was the start of the evolution of lineages and commencement of descent with modification. In other words, Darwinian evolution—in which generation follows generation and species follows species with gradual genetic change—only begins after the period of horizontal gene transfer begins to come to a close.

The fundamental mechnism by which genetic instructions in DNA are read out and converted to working proteins and enzymes (transcription/translation from nucleic acid to protein) was an intact functioning biological module—in other words, it was a self-contained, complete working process that later became a component of evolving organisms. This is a key concept in Modularity Theory. A very complex working system such as the molecular machinery that reads information in DNA to make proteins does not need to be reinvented from scratch in every organism. It is abundantly clear that in nature, once such an integrated complex process emerges, it is passed along as a complete functioning unit. This is true whether it is passed horizontally from an ancestor of a fungal cell to an ancestor of a plant cell, or whether it is passed vertically from a mother elephant to its calf. So it appears that the emergence of many extremely important biological modules predates the commencement of Darwinian evolution.

The Cambrian Explosion

The first explosion occurred at the dawn of the Cambrian period around 522 million years ago. The one-celled organisms had toiled for half a billion years to fill the world with oxygen. The ancestors of plants were now using photosynthesis to get energy by the capture of sunlight, and the ancestors of animals were now burning oxygen for energy. Multi-cellular organisms had emerged that could coordinate biological processes among a large number of specialized cells. Then, for reasons that remain obscure, all of the dozens of major animal phyla and many of the hundreds of classes and orders of animals suddenly appear on the scene, many with widely varying body plans. In the succeeding 500 million years, nothing remotely similar to this ever happens again. This explosive event has been a major focus for evo-devo scientists.

It is now clear that a major aspect of this explosion was the emergence of an embryological process called terminal addition—the repetitive process of construction of similar body components tacked onto one end of the embryo like beads on a string. Both insect segments and our own vertebrae are echoes of this event. Each of these developmental modules could be composed of a wide variety of tissues with smoothly interacting functions—muscle, nerve, skin, and so on. The various resulting body modules or groups of modules could then be assigned an individual specialization. Examples of this include the differentiation of some insect segments into wing-bearing components and some into antenna-bearing components.

The Inverted Insect and the Origin of Vertebrates

The second revolution is the abrupt and radical emergence of the "inverted insects" we call the vertebrates. One species of one of the numerous Cambrian phyla appears that is literally flipped upside down and is, at least in part, internally inverted. I say "radical" because this appears to involve a true 180-degree flip in body construction plan, and I say "abrupt" because it almost certainly took place in a single generational event. This revolution has also been a major area of focus for evo-devo.

The Dinosaur Domination

The second explosion is the appearance of a vast array of dinosaurs beginning around 235 million years ago. This now appears to be due to an abrupt change that took place in the body plan of one species of reptiles. The ancestral dinosaurs experienced a modification of their body axis—the vertebral column and surrounding structures—that suddenly altered the balance of power of life on earth, giving them the lead in energy, speed, respiration, and size. This change in the dorso-ventral (back-to-front) organization of the body

axis in development simultaneously altered all the body modules of the ancestral dinosaur to produce a sudden and overwhelming leap in effectiveness
of adaptation of several body systems all at once.

The Mammalian Takeover

The third explosion is the diversification of mammals 70 million years ago that led to the origin of all of the major mammalian orders that we know today. This explosion also appears to be associated with a change in the way in which the main body axis is organized. At a key point in the emergence of mammals, two modular reorganizations of the body axis occurred: One matched the effect of the dorso-ventral change in the dinosaurs, and the other provided greater distinction in the differentiation among the body axis modules. These two changes made mammals successful competitors in a variety of environmental niches. Simply put, these changes allowed mammals to run faster and to have more energy than the dinosaurs and lizards.

The Transformation That Launched the Human Lineage

The third—and, for many people, the most important—revolution is the body plan change that set a species of primates on the path toward becoming human 20 million years ago. What actually happened at this time is not what the scientific orthodoxy has taught over the years. The data presented in this book lead inevitably to the conclusion that the key initial event in human origins was similar to the key event that drove the dinosaur explosion and the mammalian explosion. Once again it was a sudden dorso-ventral transformation of the modules of the body axis—a re-markable reorganization of the vertebral column in which some parts moved forward and others shifted backward—that generated an upright species. In fact, the evidence shows that this upright species—most likely a ground walker—was the ancestor of both the great apes and man.

The fact that this astonishing reorganization of the primate body design in the human ancestor has escaped the understanding of thousands of scien-tists across hundreds of years does not diminish its profound importance. The major plane that separates the front part of the body from the back part changes from its standard primate position (in front of the lumbar spinal canal) to a revolutionary position *behind* the spinal canal. The first fossil evidence of the transformation was discovered more than 40 years ago (Walker and Rose 1968). A new fossil discovery made in 2004 (Moya Sola et al. 2004) confirms this remarkable event and will now force a dramatic reinterpretation of human evolution. In short, understanding the way in which major modular transformations impact subsequent evolutionary events is essential to understanding our own special history as part of a unique lineage of upright primates.

Darwin, the Modern Synthesis, and the Arrival of Modularity Theory

Certainly there are various other revolutions and explosions in the History of Life, and a number of these will be discussed in this book for context and comparison. However it is these six major events that are critical to our understanding of our own existence, as well as our understanding of how life evolves. If these six events are to be fully understood, however, it is my contention that Darwinian Theory and the Modern Synthesis of the 1940s (the addition of population genetics to the original theory) must be revised and expanded to incorporate the modern molecular genetics of biological construction.

We can safely say that Darwin simply specified descent with modification (gradual change of a species due to shifts in the genetic makeup of the total pool of genetic varieties from generation to generation), and that the various revolutions and explosions fall within this dictum. For example, Darwinian Theory and the Modern Synthesis really do an excellent job of explaining how and why the shape of the beak of a group of finches slowly changes over time. There is variation (slight differences) in beak shape among the population (the total interbreeding group) of individuals in a species. There are also various environments, and changes within these environments. Natural selection, or "survival of the fittest," will "prefer" some variants of the species over others, and gradually the gene pool of the species will change. Speciation events (the origination of a new, separate species) will wall off one group of individuals from another. If each has a different gene pool, then the two descendant species will appear different from each other, and each will be optimized for survival in the environment in which it lives.

It was 200 years ago that the great French zoologist Étienne Geoffroy Saint-Hilaire published the fantastical proposal that vertebrate animals were flipped-over versions of invertebrates. Between 1818 and 1996, no respectable biologist took this proposal seriously. Now we know it is fact. The problem for Evolutionary Theory is that this critical event did not appear to involve variation, populations, or even the accumulation of gradual (small, incremental, generation-by-generation) changes due to natural selection. Because the invertebrate ancestors remained far more prolific and varied, it is difficult to argue that this body inversion event provided improved survival for these ancestors of the vertebrates. In fact, there is a strong case to be made that what happened was a freak or "monstrous" event. It was sudden; it was lucky to have occurred in a way that allowed for it to be passed along into descendants; and it was lucky to have produced a type of creature that was at least capable of surviving to reproduce.

Of course, luck implies chance and chance implies randomness. Actually, although this revolutionary event suggests randomness from the point of view of natural selection, it is rigidly ordered and formal from the point of view of developmental biology and functional morphology (the study of how body components actually work in the life of the animal). The ancestral vertebrate was no monster, because it had an altered but complete and properly functioning developmental embryological sequence—in other words, unlike a conjoined twin or a mutated stillborn with no head or mouth, this organism had a full, working set of body components. It did not survive simply from luck, but rather because it had a novel design that met the basic functional requirements for sustained existence—namely, the capability of obtaining and processing food, avoiding predation, and successfully reproducing. There has been a fear in Evolutionary Theory that if natural selection does not act, then evolution will be seen by its opponents to have progressed at random. However, what actually seems to occur is that some steps in evolutionary history are guided by a different mechanism than the standard of population variation, selection, and speciation. It is not that evolution is progressing under no direction; rather, it is that the Modern Synthesis mechanism is not the prime player in some key events.

Why Are the Mechanisms of These Grand Events Unresolved?

By now it's probably fairly obvious that these three revolutions and three explosions are significant. But how have they been considered in the past? The first revolution—the origin of descent—simply was not known about or convincingly demonstrated until the past five or 10 years. It only became apparent as our understanding of molecular biology progressed. Though evidence of the Cambrian explosion has been accumulating for 200 years, its existence has been questioned in two ways: One suggestion is that all we are seeing is a change in fossilization—all the different types of animals emerged gradually over eons, but their fossils appear all at once because of an environmental change. (Recent molecular biology tends to support an abrupt event, however.) Another suggestion is that, because it all could have taken 1 million years or more, it was really a stately and gradual event—an evolution and not a revolution. However, when we look at the issue of remarkable generation of new body plans, it becomes apparent that even if it took 1 million years, it was a very different million years than the other 999 1-million-year periods in the billion-year History of Life on this planet.

Almost all biologists believed that the origin of vertebrates was a gradual selective process similar to the origin of any other group. Then, in one of the great ironic events in the history of science, data has recently come pouring in from the molecular analysis of the genetics of embryo formation confirming Geoffroy's theory of "flipped-over" vertebrates.

The basis for the success and diversification of dinosaurs was not understood until very recently; a similar situation prevailed for mammals. Neither of these events—the diversification of dinosaurs and the diversification of mammals—seems to have been nearly as rapid as the Cambrian explosion. If you assume gradualist Darwinian action, there is a concept of a "crown radiation"—an extensive branching into numerous different types of species in a kind of animal (for example the mammals) that has only existed in the form of limited numbers of species in the past.

An unexplained vast extinction opens up an array of environments—think of the dinosaurs dying out and making way for the mammals. At the moment of extinction for one group, a new and entirely different type of animal appears with improved features, and this group then generates a wide array of new species that become optimized for the various niches in the new environment. This argument is somewhat plausible for the dinosaur diversification, but it is difficult to explain why the dinosaurs did not come roaring back 70 million years ago (when instead the mammals became the dominant large land animals).

I am suggesting that some kinds of mutations can have such a large number of far-reaching positive effects that sudden biotic shifts—major alterations in the kinds of animals that fill the various environmental niches of the world—can be explained. Natural selection plays a role, but newly understood genetic mechanisms can rapidly give one group of animals a large advantage over another. In essence, through modular mechanisms, a single spot mutation in a critical gene can cause a remarkably large number of changes in the resulting adult. For example, a single point mutation in a gene that controls the shape of the connection between rib and vertebra leads to an alteration of *every* body segment, which immediately achieves a major improvement in the mechanics of breathing, and alters the way the animal's body moves during walking and running.

Finally we come to the case of human origins. Since the pioneering work of Linnaeus and of the German poet Goethe in the 1700s, it has been clear that human origins are explainable by the same forces and processes of ordinary biology that have led to the origin of every other species. However, there is ample evidence that a revolutionary event did indeed take place that set our lineage of upright primates on its course.

The event in question is similar to the type of change that occurred in the case of the origins of the vertebrates and the explosions of the dinosaurs and of the mammals. In an ancestral ape, a critical change took place in the distribution of the position of anatomical structures from front to back of the animal (dorso-ventral patterning), affecting various structures arrayed along the main anterior-posterior (head-to-toe) course of the vertebral column and associated structures (in science-speak, the "embryological longitudinal body axis"). This change appears to have produced the anatomical basis for our upright posture in a sudden, single event. Recent and definitive fossil evidence now shows that this took place before the evolution of the unique form of our hands or our brains. Given the way in which the subsequent evolutionary changes occurred after the initial, major change (that is, given its "downstream effects" on evolutionary context), there is a powerful case that this is what established our upright lineage and what set our ancestors on the path to our current state of existence.

Although the initial fossil evidence for this event was discovered 40 years ago by Alan Walker and Michael Rose (Walker and Rose 1968), new discoveries in the past few years have confirmed this startling proposal and will now lead this theory to displace the ideas accepted and taught by virtually all of the academic specialists in this field. Taking an entire generation of specialists in human evolution and proving them all to be wrong about what they learned from their own professors, as well as what they have taught and believed for their entire careers, will not take place without a great deal of controversy, protest, and denial. Nonetheless, the facts speak for themselves, and the truth on this point will prevail.

Objectives of the Book

Within the field of evolutionary science there has been, and continues to be, enormous turmoil. Biologists committed to Evolutionary Theory are forever "circling the wagons" to fend off attacks from poorly educated creationists or advocates of "intelligent design." Even within the circle itself, there is enormous turmoil. In questioning and rejecting scientific orthodoxy, no mass of credentials will convince a spurned scientist that he or she should give way and accept that they have spent a career believing, teaching, and publishing in error. Certainly, it is equally unlikely that the advocate of an erroneous religious dogma will realize the error of his or her ways. Nonetheless, the purpose of this book is to call into question some key elements of what has long been said and taught about evolution in general, and about human origins in specific. Along the way, recent reappraisals in philosophy, biochemisty, theology, and zoology will be identified, and a number of traditional concepts whose time may have past will be identified and updated—regardless of the potential for controversy and hurt feelings. As the reader, you are invited to learn and follow. All that is required is a small measure of patience, an open mind, and a spirit of shared inquiry.

History of Evolutionary Theory

In the face of a growing mountain of scientific information about natural biological species, geologically stratified series of fossils, and anatomical evidence of unity among living things, Darwin proposed his Theory of Evolution in 1859. He supported transmutation, or the potential for a new species to arise as a descendant of a parent species. He advocated descent with modification, in which inheritable changes that distinguish an offspring from its parent are the raw material for evolutionary change. And finally, he believed that natural selection ("survival of the fittest"), acting upon variation among individuals, would gradually change an ancestral species into a descendant species that was altered in outward form.

The basic laws of genetic variation were worked out and published by Gregor Mendel in 1866. Once Mendel's work had been "rediscovered" in 1900, a concrete basis for the inheritance of traits seemed to be understandable. In the 1920s—long before we understood DNA—Ronald A. Fisher applied statistics to the problem of variation of genetic traits among the individuals of a species (a population). In the following years, Theodosius Dobzhansky and others were able to show how this understanding of patterns of inheritance supported Darwin's ideas. The fusion between Darwin's theory and the population genetics of the 1940s led to the "Modern Synthesis" combining those two disciplines, and this fusion is what we generally refer to as Evolutionary Theory.

Finally, in the decades following the 1940s, the structure of DNA was discovered and described by James Watson and Francis Crick (Watson and Crick 1953), and a vast array of other discoveries in molecular genetics was accommodated into the Modern Synthesis.

Genetic Basis of Embryological Development

Beginning in the mid-1980s, the genetic code that described the construction plan for animals and plants began to be decoded. As this grand project has proceeded, things have not gone smoothly for the Modern Synthesis. In many ways, the more we understand about the way that organisms are actually assembled, the more we see that Darwinian Theory and the Modern Synthesis are inadequate to explain some of the most important events in evolution. Using astronomy as an analogy, it is as if we had a science that explained how planets originate and how the galaxies of the universe were organized, but that could not explain the beginning or potential end of the universe, or the creation or death of a galaxy.

The seemingly endless stream of breakthroughs in morphological genetics (the exploration of the genes that control the construction of an organism during its embryological development) during the past 20 years has had a peculiar relationship to Evolutionary Theory. A very odd thing has happened. You see, we have a sort of junk heap of failed ideas and theories about evolution. There are a number of long-rejected major theories that preceded Darwin's successful proposals, and, since Darwin, there have been numerous biologists—mostly discredited and ignored—who have identified problems with both Darwinian Theory and the Modern Synthesis and proposed alternative views. What has happened is that in many cases, now that we have the hard data from morphologic genetics, we see that the discredited ideas and theories often seem to be correct after all!

Even more disconcerting is the fact that morphologic genetics is not really a field of theory per se. It is hard objective science, consisting of gene sequences, laboratory data, and endless amounts of supporting detail accumulating at an increasingly dramatic rate. Even the most ardent supporters of the Modern Synthesis of the 1940s understand that things are about to change significantly in the field of evolutionary biology.

The most important objections to current Evolutionary Theory are as follows: Although it does a great job explaining how populations of organisms gradually change in response to the environment, it does a poor job of explaining constraints on anatomical change—or, to put it another way, why so much similarity still remains. (Why, for example, should there be so much similarity between the number of bones in the limb of a frog and a human?) It also does a poor job of explaining discontinuities, or how sudden major redesigns occur, such as the emergence of a new "body plan" when a new phylum or order of animals first appears. Additionally, it provides no basis for assessing hierarchy in the progress of complexity (that is, it cannot say why some types of changes in design are more important than others). And finally, it cannot explain directional trends, or why some mutations lead to a series of related and consequential changes independent of the accountable effects of natural selection, effects such as the emergence of a wide variety of numbers of body segments in animal groups whose ancestor became capable of adding additional segments.

Each of these four phenomena—constraints on change, discontinuities, hierarchy, and direction—is critical to a proper understanding of how and why humans evolved. The failure of Darwinian Evolutionary Theory and the Modern Synthesis to provide strong and comprehensive explanations in these areas has led to endless series of attacks from those who question the role of biology in human evolution altogether. Some writers, such as Richard Dawkins, have recently pointed out that existing Darwinian theory is broad enough that it can't be disrupted by these issues (Dawkins 1996), and Ernst Mayr (Mayr and Provine 1998, Mayr 2001) has delighted in skewering one conflicting proposal after another. (As graduate students we took great pleasure in feeding him comments heard earlier in the day in our course with Stephen Gould.)

The sudden emergence of the vertebrate body plan in an abrupt, 180-degree flip can be said to fit into the standard Darwinian framework because it is still "descent with modification." The fact is that gradual change cannot be pinned down to a particular time frame—we can still call the evolution of vertebrates "gradual," even if it took only hours for the momentous change to occur instead of millions of years. However, there is a great difference between a body of scientific ideas that has to be unreasonably stretched and distorted by rhetorical flourishes so that it can accommodate new facts, and a better theory that robustly explains these new facts.

The principal case for change is not necessarily the accumulation of 65 years of modern molecular data since the assembly of the Modern Synthesis. Instead, it is the fact that a growing segment of this data tends to support the ideas of Goethe and Geoffroy, both of whom wrote 65 years before Darwin published *The Origin of Species*.

History, Philosophy, and Science

Although this book examines molecular genetics, it also explores the role of history and philosophy in guiding the ideas of scientists. There is no escaping the importance of these subjective factors and how they have influenced the history of scientific inquiry and thought. For example, the choice to emphasize the study of vertebrates—and humans in particular—at the expense of invertebrates, is not entirely an objective choice. Scientists are always influenced by external cultural constraints and priorities.

We all understand that it is appropriate for the medical scientist to emphasize the study of areas of biology that can help lead to the prevention or treatment of disease in humans, and that the greatest attention should be directed to the diseases that cause the most harm. But what analogous priorities can help us choose where to focus our efforts in the study of organisms? Are those priorities rooted merely in what is "interesting," or do they grow out of the dictates of philosophy, history, and/or politics?

It is easy to apply history and philosophy to show why we care far more about *Homo sapiens* than about 100,000 different species of algae. We can also use history and philosophy to help us reject the idea of abandoning the study of humans in favor of spending all biological funds on researching unknown insect species. Beyond that, we can apply history and philosophy to help us understand why scientists working in a particular cultural context come to understand their findings in particular ways. The point is that science needs to be objective and it needs to be accurate, but it does not and cannot exist in a cultural vacuum.

Structure and Function, Plato and Aristotle

The current ideological struggle in biological science is often—and somewhat inaccurately—portrayed as a recurring difference in philosophy between *typology* (the search for an idealized symbolic representation of the essence of an animal) and *functionalism* (the search for an explanation of the purpose of every unique feature of an animal species). The "evo-devo" enthusiasts are accused of a semi-religious adherence to a notion of ideal types of animals, types that are epitomized by one of a small number of standard patterns of embryological development. The "adaptationists"—so named because they believe that every feature of an animal is optimally adapted to its function through the action of natural selection—are accused of telling "just-so" stories or inventing rationalizations about how wonderfully a specific anatomical feature is fine-tuned for its necessary use. Most likely this is done in hopes of explaining the purpose of every feature in an organism, even when logic and biology show that some features exist due to their history or role in development and have no actual functional effect on survival.

Stephen J. Gould provided an excellent summary of the issue in his famous story about the spandrels of San Marco (Gould and Lewontin 1979), as well as in his recent reply to the subsequent attacks on this analogy (Gould 1997). The story pertains to an architectural feature of some of the arches in the walls of the great Cathedral of San Marco in Venice. The spandrels, or spaces between the arches and their rectangular enclosures, feature important mosaics. Were the mosaics put in the spaces because the architecture happened to result in the construction of many little alcoves ideal for statues? Or were arches with spandrels chosen because the cathedral had to have spaces to show off mosaics? Gould argues that the spandrels, as is the case with many morphological features, are byproducts of the main architectural requirements of the assembly, and only seem to exist because of whatever function they happen to perform.

When we observe a structure in an animal that appears to be active in a particular function, how can we be sure whether it is adapted to that function or whether it is just an echo of the embryological formation processes and history of the organism that happens to have some apparent function? Are all the dozens of bones of the skull present because each has a specific function, or are they present because that is the essence of what nature employs in the assembly of all skulls? All of this is, of course, just a recasting of the classic and ancient dichotomy of philosophy. Plato—as do the evo-devo enthusiasts—taught the search for ideals, while Aristotle—as do the adaptionists—believed each individual component of the universe should be described and its function determined.

For our purposes, it might be helpful to recall the most widely recounted bit of Plato's writings, the allegory of the cave (Plato 1991). In the cave, we have a group of prisoners chained to the ground from a very young age, and unable to see their surroundings. Their gaze is permanently fixed on the cave wall in front of them. Behind them is a tall curtain, behind the curtain is a platform, and behind the platform is a fire. On the platform, men carry various artifacts above their head—small statues of people, animals, plants, and objects—and the fire causes shadows of these artifacts to appear on the cave wall. The moving shadows that the prisoners see is their entire world, so they develop elaborate explanations of what the shadows are, and how and why they move.

Plato asks us to consider freeing one of these prisoners so that he can see that the shadows are mere projections of actual objects being carried by people. He is brought outside the cave to see real plants, people, and animals, and finally to see the sun as the true source of light. Plato's allegory can be seen as a description of the mission of science and philosophy to pursue true and essential knowledge.

Aristotle came from a long line of physicians and had an interest and background in natural science as well as philosophy. Although we do not know entirely how Aristotle viewed Plato's allegory, it seems that Aristotle's view would have been to describe everything: the sun, the animals, the cave, the artifacts, the fire, the prisoners, the shadows. You now have a complete description, you suggest functions for everything, you write it all down, and you move on to the next subject matter for description. Aristotle's works in biology and anatomy represented the high point of knowledge in these disciplines for more than 2,000 years.

Aristotle's zoological works are often described as a baffling series of detailed descriptions of the wide variety of animals he encountered. Although he proposes major categories for grouping organisms, he does not appear to be greatly concerned with deciding what aspects of an organism are most important to study. He just wants to describe the various organisms he has encountered, write down what he has noticed, suggest functions, and move on. This is not all that different from the approach of the modern cladists described earlier. Additionally, Aristotle expects to be able to describe a definitive function for every anatomical feature he describes, which is an essential basis of Darwin's view.

The modern evo-devo school, on the other hand, embraces the Platonic approach. They look for the true underlying patterns through which we can map out an explanation of the assembly process of all animals. If we can fully explain the basic process, it is not that important to describe every minor variation in embryological development that produces the various differences among the millions of different species of organisms.

Goethe, Geoffroy, and Lavoisier

The first major modern work to apply an idealistic or typological perspective to the biological world came from the German poet-naturalist Johann Wolfgang von Goethe. Goethe approached biology to look for an ideal type of plant, and his writings on the subject are typically portrayed as the source and prime example of this kind of thinking. However, I believe that Goethe had a second idea about biology, one that bridges the intellectual space between the functionalist and typological views. This is the concept of *modularity*, in which biological entities are composed of repeating, essential units.

We now understand that modularity exists at all levels in biology, from the biochemistry of enzymes to the organization of animal body plans. Understanding modularity in biology is proving to be the key to bridging the gap between functional and structural views of biology. Truly, the great contribution of molecular genetics to biology is that it has enabled scientists to understand the underlying composition of these biological modules. Improved understanding of modularity is the driving force behind the pressure for change in Evolutionary Theory.

Both Goethe and Geoffroy seem to have gained their inspiration for a modular approach to biology from the spectacular, pioneering chemical work of Antoine Lavoisier. Aristotle had written that all of matter was composed of four elements—earth, air, water, and fire. But it was Lavoisier who disrupted this by going on to identify 33 elements, such as oxygen, nitrogen, hydrogen, mercury, zinc, sulfur, and phosphorus. Lavoisier's list of elements was a key predecessor to the periodic table of elements.

Goethe started as a writer, but eventually became a Weimar bureaucrat. It seems odd and arbitrary that he accepted a position as Minister of the Department of Mines. However, it was in this position that he participated in the first International Society of Mining Science in 1786, a small group that included Lavoisier. Within months of learning from Lavoisier about his triumphant accomplishment, Goethe quit his job in Weimar and left Germany to start his search for the fundamental composition of biological life.

Geoffroy was 20 years old when he met Lavoisier at a series of small scientific meetings near Paris. In the following year, in 1793, Geoffroy had become the first professor of zoology in Paris, and started his first research into a unitary plan of components that could be detected in any mammal. Lavoisier was executed on the guillotine in 1794 during the French Revolution—primarily because of his aristocratic origins and his involvement in a tax collection business. ("The Republic has no need of geniuses," said the judge.) However, Lavoisier's scientific collaborators—Laplace and Bertholet—both spent considerable time with Geoffroy later in the decade while he was further developing his zoological ideas.

Lavoisier of course did not develop his chemistry in isolation. There were other chemists at work on the great problems of the age, but Lavoisier's quantitative experiments were often decisive. He studied the decomposition of water into two gasses—hydrogen and oxygen—and their recombination to produce water. He showed that air contained both oxygen and nitrogen, and that both fire and the respiration of animals involved the combustion of oxygen. By the late 1780s, Lavoisier was starting to write his famous textbooks of chemistry and was greatly admired as one of the great geniuses of the century. This was the period of time when both Goethe and Geoffroy learned of his accomplishment. How tempting it must have been to these two men to try their own hands at overturning the ancient wisdom of Aristotle!

The impact of timing and chance meetings, and their inspirational effects among scientists, cannot be underestimated. Lavoisier had the tools to succeed in his experiments with the chemical elements, but the vision of Goethe and Geoffroy could not be adequately explored until the 20th century, when molecular biology finally enabled us to detect the actual molecular basis of the modules that these two men envisioned so many years before.

A Modularity Theory for Evolution

Any thoughtful observer will immediately appreciate that there are similarities among the creatures of the earth. A giraffe and a mouse and a human all have one head, two eyes, one mouth, one tongue, teeth, lips, four limbs, and so on. In fact, at increasingly minute levels of detail, the presence of similarity continues: Each has seven vertebrae in the neck, 12 ribs, two lungs, and a heart with four chambers. The similarities continue even to the level of cells and proteins, eventually encompassing thousands upon thousands of features designed in essentially similar fashion.

When confronted with an extraordinary recurring pattern such as this, the human mind is impelled to ask how, why, and what. What is the basis for this similarity among organisms? Modern science teaches Darwin's view that similarity among organisms arises because they have descended from common ancestors. Orthodox theology says simply that God chose to make creatures this way. The ancient Greeks believed that the perfect human form was modeled upon the form of the gods, with all other creatures being degradations from the human god-like form while still retaining various similarities to us. A similar line of thought was also embraced by later Christian theologians, who took literally the biblical dictum that God made man in his own image.

It was in the late 1700s, three generations before publication of Darwin's famous work, that science first began to contend with theology and philosophy over this issue. The first great scientific explanation for these similarities was inspired by the anatomy of the spine, and it came from the great German poet Johann Wolfgang von Goethe (pronounced "gur-tah"). Not genes, and not ancestors, but vertebrae seemed to hold the key. His ideas were further developed in elaborate and formal detail by the very colorful and very inspired French biologist Étienne Geoffroy Saint-Hilaire. By 1850, this concept of looking to the vertebra as a central unifying aspect of biology was a fixture of textbooks and widely accepted throughout the world.

Eventually, cracks and flaws in the vertebral theories began to appear in the 1830s. After sustaining a series of attacks over several decades, the publication of Darwin's *The Origin of Species* in 1859 seemingly laid it all to rest. Ironically, however, discoveries in genetics at the close of the 20th century have begun to show that the vertebral theories were indeed far closer to the truth than the Darwinians ever imagined.

So much has changed that a new, more comprehensive theory of evolution is called for. This new theory can trace its origins to the concept of the "protean" component (from Proteus, a Greek god who could appear in an infinite variety of forms) in Goethe's biological writings. This underlying module or monomer can be multiplied and modified in a myriad of ways to create an endless array of biological species. For Goethe, the vertebra symbolized the protean potential of biology as it applied to animals. In modern terms, his concept has been further clarified and recast as the segment-organizing potential of the *Hox* genes—the genes that help direct the formation of the vertebrae and other repeating (or *homeotic*) large-scale structures in animals.

Goethe's Idea

The arrival of Goethe's first archetypal proposal on the intellectual scene had an electrifying impact upon the biologists of his time. Here was an audacious poet and playwright who sought to turn the world of science upside down! The archetypal idea was strong and appealing, and the fundamental observation underlying it was compelling. As I will explain in Chapter 2, Goethe had not yet fully formulated his "Proteus" concept when the scientific world learned of his archetype. Although he had delayed publication while continuing to ponder the underlying ideas, history forced his hand. In 1807, Lorenz

Oken announced and published a vertebral theory of the skull that closely matched the ideas Goethe had been discussing, without publication, for the past 17 years. Also in 1807, French biologist Étienne Geoffroy Saint-Hilaire published a work that included similar concepts, showing the detailed similarity of the skeleton of fish to the skeletons of land animals.

A Convergence of Thought

Geoffroy was drawn to the new idea by his discovery that standard parts of the skull and vertebrae that appeared to be missing in some kinds of animals could be identified as rudiments in their developing embryonic skeletons. Oken, on the other hand, was attracted to a vaguely metaphysical concept of transforming one structure into another in order to build new types of animals. And finally, Goethe was concerned with the vertebra as the fundamental monomer in animals (as opposed to the leaf as the monomer in plants). Much attention was paid to Goethe's claims of priority over Oken. However, the general consensus in the scientific and intellectual community seems to have been that all three had conceived of an archetype composed of repeating vertebral elements.

The common element of these ideas had a powerful effect in the scientific community around the world. One had only to look to the vertebrae, see the progression of repeating elements, and comprehend the relationship among them. Any scientist who did these things and knew of the idea found himself drawn to the same remarkable conclusion: These repeating elements seemed to hold some intrinsic secret, some underlying mystery, which could unify and simplify the entire world of biology.

At the same time, however, Johann Meckel, a prominent anatomist who was the editor of one the main academic journals of that era (*Archiv für Anatomie und Physiologie*), published a detailed anatomical paper concluding that the skeletal elements that make up the skull do not have a direct relationship with those that make up the spine. The vertebral archetype concept clearly grew out of the realm of philosophy and metaphysics, whereas Meckel reached his conclusion based on various formal anatomical grounds. Nonetheless, the vertebral insight provided a compelling and intellectually satisfying vision that could not really be successfully assailed just because it lacked anatomical details.

The proposal of vertebrae and "vertebralized" skull elements as a unifying structure was attractive beacuse it accommodated mammals and lizards, snakes and fish—all of the creatures of the vertebrate group. Indeed, the very subject of categorizing and naming all the animals of the earth was heavily influenced by this proposal.

Impact on Understanding the Animal Kingdom

Linnaeus and the Organization of Life

In the 1700s, Swedish botanist Carolus Linnaeus was the first new major contributor to Aristotle's descriptive approach. Linnaeus's first project, at the age of 25, was to explore Lapland and describe its plants. Three years later, in 1735, he published the first edition of his *Systema Naturae* as an 11-page list of species. By the time it reached its 13th edition in 1770, the work had 3,000 pages, and provided names and categorization for nearly 20,000 species.

Linnaeus began applying Latin binomial names (such as *Homo sapiens*) to plants and animals. The terms genus and species were originally general aspects of Aristotelian logic that were borrowed for the naming of organisms. In the Linnaean system, the two types of name are used to link similar types of animals. As he gradually built his lists of species, he came to understand that genera could be grouped together based on more fundamental and hence progressively wider aspects of similarity. For instance, it seemed reasonable to list all the different species of monkeys in one group, all the fish in one group, and so on.

To provide structure to the organization, he applied subdivisions at various levels, echoing the various levels of an outline. Going from lesser to greater specificity, the terms included kingdom, phylum, order, class, and family. These terms seemed to describe an organized structure in the natural world, but whether this apparent hierarchical structure is a real biological phenomenon or simply an arbitrary set of terms imposed by the naming system is still a matter of active debate among biologists, and is a significant subject of this book.

Linnaeus put cattle and sheep in the Bovidae family group, dogs and wolves in the Canidae family, and cats and lions in the Felidae, then put all of these together as mammals, and so on. By identifying certain major common characteristics, larger and more general groupings could be identified. The concept of mammals having fur or hair, being warm-blooded, and feeding their young with milk from the breast was one such example.

The work was more than just a list, however. As information accumulated, he reclassified whales as mammals rather than fish. This sort of revision anticipated the modern view that the naming and organizing of animal groups also involves a process of trying to determine their underlying relatedness, similarity, or history. Perhaps most importantly, Linnaeus departed from those who went before him by including humans along with the monkeys and apes in the order Primates, rather than apart from the rest of the animal kingdom. This work is important in that it is in continuity with ongoing research that continues to this day. It also provided one of the major areas of context in which Darwinian Theory emerged in the following century.

Cuvier and the Basis of Animal Design

Georges Cuvier, a colleague of Geoffroy, believed that Linnaeus had tended to over-generalize the many differences that distinguished the various types of animals. In his seminal work published in 1812, Cuvier identified four major groups: the Vertebrata, the Articulata (insects), Radiata (jellyfish), and the Mollusca (clams). At the very least, these divisions were meant to demonstrate four fundamentally distinct designs for animals.

Geoffroy, however, worked in the opposite direction. Throughout his career, he developed methods and collected material to show that a single type of underlying design or plan could be identified in nearly all types of animals. Through Geoffroy's linkage of his own ideas to vertebrae, the concepts of Goethe's archetype and his later views about modules could be brought into play for a truly vast array of creatures.

Goethe's idea by itself provided a controversial explanation for the phenomenon of similarity among animals. It entered an effective vacuum in that there was no other compelling scientific explanation for what biologists had observed in comparative anatomy. Geoffroy's elaboration of this idea, however, called for rejecting the conclusions of Cuvier, who was one of the leading biologists of the time. Worse still, Cuvier felt that Geoffroy's claim was transparently based on philosophy and metaphysics, and not upon scientific observations.

Although Geoffroy's classification scheme was widely abandoned after 1830, Goethe's basic insight survived intact. In fact, the concept of an anatomical archetype became the orthodox scientific view at the time. Its greatest proponent was Richard Owen, a prolific anatomist who generated some of the great biological works of the mid-19th century.

Owen, the Archetype, and Homology

Owen's great mission was to simplify and unify the naming of the anatomical parts of the body. Prior to his time, it was common for the upper arm bone of a horse, a dog, and a human to have different names, but he introduced a uniform naming scheme in which the bone would be called the humerus in all three creatures. This extended to a minute level of detail, including, for example, the names for the small bony projections on the surface of the skull and vertebrae. Even more important, Owen tackled difficult problems such as correctly naming the bone inside the hoof of a horse by showing that it was fundamentally identical to the last bone of the third finger in a human.

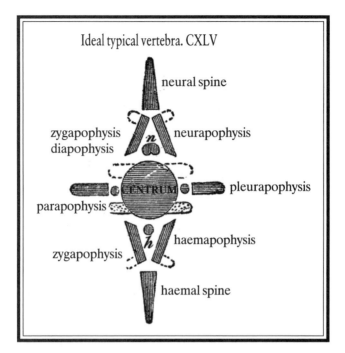

Figure 1-1: Sir Richard Owen's ideal vertebra. Owen's fundamental "animal crystal" depicted the ideal as a radially symmetric object with a set of extensions in the four directions of the compass. This structure was then repeated down the length of the animal to form the archetype.

To provide a conceptual underpinning for his naming project, Owen developed the concept of *homology*. This term denotes a fundamental similarity between structures in two different animals, even if they do not appear to be entirely similar. The concept of the archetype served Owen well in this project because it provided a theoretical basis for his proposals of homology. According to this reasoning, both the hoof and the finger were derived from the same component of the archetypal design (see Figure 1-2).

Owen was also deeply involved in paleontology. Based on fossils records, he reported in his publications the progression of some species across time. Among the most famous of these is his revelation to Darwin that a giant fossil creature found in South America during the voyage of the *Beagle* was a huge sloth related to the small modern sloths in the region. The fossil species were discovered in a series of *geological strata*—layers of ancient earth and rock stacked on top of each other in a hillside cut open by a stream or earthquake. The species from strata progressively closer to the modern earth's surface looked progressively more similar in appearance to the modern animals. This was a key event in the intellectual development of the young Charles Darwin.

It may seem strange to the modern mind, but Owen considered these stages of development to be evidence of serial independent acts of creation by God. He was most interested in using series to show that he had correctly homologized structures such as the horse's hoof with the toenail of the middle digit of horse-like species found in deeper geological strata. Unlike Goethe, Owen saw the archetype as a direct manifestation of the hand of God. He wrote and lectured about animal creation as evidence of God's work (Owen 1864). In the 1830s Owen and Darwin were scientific collaborators, but later— perhaps predictably—they became bitter antagonists. If there was one man who represented scientific and theological orthodoxy at the time of the publication of Darwin's *The Origin of Species*, that man was Richard Owen. Darwin's acolytes made a mission of tearing apart Owen's legacy, and he in turn made a career of attacking Darwinism in his later life.

Rejection of the Vertebral Theory

In 1855, the German embryologist Robert Remak published extensive evidence that vertebrae were not fundamental in the assembly of organisms. Essentially, he showed that, in vertebrates, there appeared to be a solid block of tissue called the mesenchyme that was divided into vertebrae late in the course of development of the embryo. This finding appeared to directly contradict the concept of vertebrae as repeating units out of which the entire organism was assembled.

At the Croonian lecture delivered to the Royal Society on June 17, 1858, Thomas Henry Huxley assembled the evidence from Remak's work in embryology of segments, as well as a variety of other evidence showing that the bones of the skull were formed in an entirely different fashion from vertebrae. Based on these points, he convinced many of the leading scientists of the time that Goethe's central idea of the vertebrae making up the fundamental constituents of an archetypal plan for the animals, although championed by Richard Owen, was philosophically interesting but scientifically invalid.

Huxley and others presented evidence that the skull was not made of modified vertebrae. The vertebrae were split out into a series of units only to serve the needs for locomotion and not as a manifestation of some grand master plan. Furthermore, insect segments had nothing to do with vertebral bones. Therefore, Huxley asserted, the idea of vertebrae providing a key to the nature of similarity among organisms was intellectually bankrupt and must be abandoned. In the following year, in 1859, Darwin published *The Origin of Species*, and Huxley embraced it with a passion. Remak had rendered the archetype vulnerable and Darwin finished it off.

In Goethe's original view, the archetype was a sort of unexplained, metaphysical starting point for organisms, with each species optimized and differentiated to fulfill its own particular role in nature. Owen formalized this

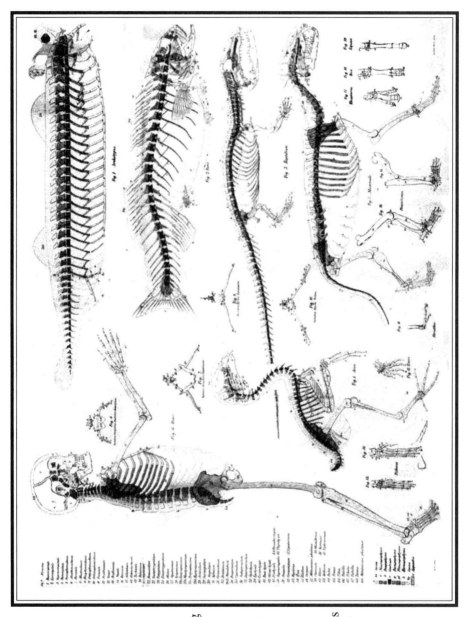

Figure 1-2: Owen's archetype and manifestations among vertebrates. Working from his "ideal vertebra," Owen showed how repetition of the vertebra and modification of its four processes could be used to construct the various types of vertebrate animals as well as the human skeleton.

under a concept of *parcelation* (in which the archetypal plan was effectively the way in which the Creator began his work), with specific attributes being modified to suit the intended activities of the species being created. In Darwin's model, however, the role of the archetype as the source of the plan was taken over by the role of the common ancestor, and the process of modifying the attributes of a given species was called "adaptation through natural selection." Darwin also departed radically from Owen in that he did not see species as fixed or static. New species could arise at any time, and the anatomy (or *morphology*) could change gradually as one species split off from its parent species.

From the publication of Darwin's great work in 1859, the race was on to understand how adaptation worked and to learn the way in which it interacted with the essential biology of common descent. Naturalists studied the transformation of species, and 20th-century geneticists studied how genes changed in natural populations of animals. It became apparent that evolutionists could learn far more about the process of change in animal shape by studying the beaks of birds or the shells of snails than they could by studying the vertebral bones of the spine. In the 12 decades that followed the publication of Darwin's *Origin*, the study of the spine was relegated to an area of minor anatomical interest among biologists. A very large body of medical work about the spine in back pain and in injury developed and grew, but there was no important role for vertebrae in biology.

Modern Science Confirms the Vision of Goethe and Geoffroy

Then, 120 years later, in 1979, something remarkable happened. Using scanning electron microscopy to re-explore the early events in embryology, two anatomists, Patrick Tam and Stephen Meier, at the University of Texas in Austin, found Goethe's cranial vertebrae (Meier 1979, Meier and Tam 1982). They discovered what they called *somitomeres*—the basis for the development of segments in both the skull and in the vertebrae—forming very early on in the embryological development of mouse embryos. The exact basis for and genetic details of the somitomeres is still being explored. However, from this and other discoveries, it began to seem as if Goethe and Geoffroy had been right after all, and Huxley was wrong.

Even more astonishing was a discovery from the field of genetics published five years later. In 1984, William McGinnis and his colleagues, working at the University of Basel in Switzerland, found the genetic instructions that direct the early assembly of the body segments in the fruit fly *Drosophila* (McGinnis et al. 1984a,b,c). A technical term for segmentation is *homeosis*, so they called the DNA sequence a *homeotic* gene. Using the DNA spelling of the insect *homeobox* sequence, found in several fruit fly homeotic genes as a basis for comparison, Charles Hart and his colleagues at Yale discovered in 1985 that the genetic instructions for segments in mice and in humans relied on the same homeobox gene sequence as in insects (Hart et al. 1985).

Geoffroy's remarkable claim for unity of plan had been right all along. There is a fundamental similarity between insect segments and the vertebrae of the spine. In 2003, we learned that Geoffroy was correct in his even more outrageous claim—namely, that the deployment of *Hox* genes in the squid is consistent with their axial sequence in other animals (Lee et al. 2003)—that led to his academic demolition in his debate with Cuvier in 1830 (Appel 1987). Goethe's original insight from poetry and philosophy had been correct, and 200 years of accumulation of scientific detail had only served to obscure the truth!

From the Antennae of Fruit Flies

The study of homeotic change in fruit flies was the original source of many of these revolutionary findings. Fruit flies are studied by biologists because their lives are very short, and because so many of their features can be seen by examining them under a low-power microscope. Each segment of a fruit fly has a specific function. The first segment has eyes, the second has antennae, further along the body come segments with legs, then wings, legs, and so on.

Occasionally in nature, a mutation occurs and a malformed fruit fly is born. Mutations that affect the correct determination of segments are called *homeotic* mutations. The most famous of these mutations is called *Antennapedia* (see Figure 1-3). These are fruit flies that are born with legs instead of antennae sprouting from the head segment (or vice versa, for the "null" or non-funtioning version of the gene). Other mutations include *ophthalmoptera*, in which wings appear in place of eyes, and *bithorax*, in which an extra pair of wings occurs. In these mutations, the segments of the insect form, but the segment gets the wrong instruction as to what it is supposed to grow into.

In the late 1970s, one such mutation occurred in an individual fruit fly, a homeotic mutation transforming leg to antenna in *Drosophila*. Gary Struhl at the University of Cambridge understood the significance and reported the genetic mechanisms in the journal *Nature* (Struhl 1981). This mutation was studied and evaluated, and the fly's descendants were studied. More mutations of a similar type were sought out, and eventually the secrets of the genetic instruction plan for the assembly of animals came to be known. By comparing the genes of normal flies with the genes of homeotic mutants, the site of the mutation was eventually located. Later, when it became possible to decode the DNA, the exact genetic spelling for several of these homeotic genes was learned. Each of them contained a sequence that apparently tags them with instructions about which segment to build. That tagging sequence is called the *homeobox*.

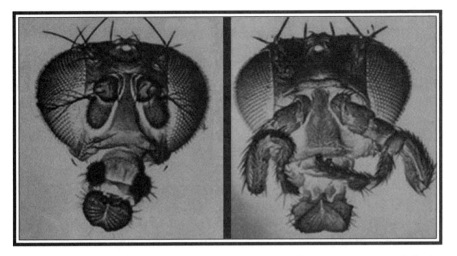

Figure 1-3: The Antennapedia homeotic mutant. An important source of the key breakthrough that has ultimately led to the deciphering of the genetic code that spells out the embryologic assembly instructions of most animals. This mutant fly has fully formed legs in place of antennas. It is a homeotic error mutant because a structure has been placed in the wrong body segment (Griffiths et al. 1996).

The way in which animals are assembled is also studied at the actual physical, anatomical level. During the 1800s and much of the 1900s, the most important tool for studying early development was the light microscope. However, much of what the microscope could teach us about the early events in embryological development was already know by the end of the 1800s. The introduction of the scanning electron microscope (SEM) in the 1960s launched a reexamination of what had been learned in the last century, and it was this work that led to the discovery of the somitomeres by Tam and Meier in 1982.

The SEM can provide remarkably dramatic, three-dimensional views of very small structures. It was SEM images of early mouse embryos that revealed the early segments or somitomeres missed by the 19th-century anatomists. Not only could the somitomere segments be seen lined up through the head and body at a very early stage, but their developmental fate could be followed. In this fashion, it was possible to show that the components of the skull as well as the various vertebrae were all formed by modifying the particular shape of a common source unit. These common repeating units, the somitomeres, are exactly what Goethe's vision called for.

Because the anatomists of the 1800s did not have the proper equipment to observe these repeating units of the early embryo, they could not accept that they might actually exist. After all, in science, observation is king. Nonetheless, observation is directed at testing theories, and science needs theories and ideas to direct and guide it. There is an awesome leap of faith required to generate an entirely new theory that departs from the way everyone else sees nature. Scientists may find their whole careers turned upside down or banished into insignificance by the arrival of a new theory. Of course, the theoretician with the new idea is always challenged to provide evidence for it. Established scientists who feel threatened by the new theory often try to throw competing evidence against it, and an extended intellectual battle may then ensue. Scientists do tend to share a belief, however, that the truest and best theory will win in the end because the observable evidence of nature will ultimately support it.

Irony arises in science when a theory is correct, and when the supporting evidence exists, but scientists are unable to see that evidence. This can occur when the methods of observation available to scientists are not advanced enough or effective enough to allow them to observe the evidence. It might be beyond the level of detail of the best microscope, or it might involve a type of chemical that existing instruments cannot detect. This is the classic notion of an idea that is ahead of its time, which appears to have been the situation with Goethe's vision.

Do the findings of *Hox*-determined cranial segments and homeoboxes really align with the ideas of Goethe and the other advocates of the archetype? I think the answer is yes. Under pure Darwinian thought, similarity among animals has to do with their history rather than with their design. Animals look the way they do or are constructed the way they are because of what or who they are descended from. They differ from their ancestors because of what they do today or, essentially, because of who they have become. There is not enough theoretical space in classic Darwinism to accommodate the impact of common needs for the design and construction of animals. This question of the role of fundamental rules of organismal design is what Goethe was thinking about when he formulated his theory. (Remember: It was the revelations of elements in the chemical world that led Goethe to look for his repeating components underlying the design of animals.)

Stephen Gould has tried to shore up Darwinian thought in this area by proposing that one important component of the history of an organism is the set of constraints on the possible forms it can take that it inherits from its ancestors. He proposed the idea of a *morphospace*—a conceptual vast array of all possible imaginable appearances of organisms (five heads, a leg coming out of the middle of the abdomen, and so forth)—and asked why the designs of organisms fill so little of this potentially vast morphological space. He therefore proposed certain "channels" in which organisms are able to evolve their shapes.

I would tend to support Goethe's view, however, that the repeating elements in the center of the body axis are the most fundamental source of design constraints in the animal body. We still don't really know why that is the case, but it seems very convincingly clear that, to put it simply, vertebrae do matter. Evo-devo and *Hox* genetics have been criticized because these disciplines rely on detailed study of a small number of existing species in order to make inferences about vast numbers of extinct ancestral species—species that obviously cannot be studied with genetic or embryological methods. However, vertebrae can provide excellent paleontological detail that reflects the way in which body axis modules evolve. They help provide a bridge between the methods of classic evolutionary science and the methods of morphological genetics. The origin of upright posture in humans, the way that dinosaurs ran on two feet, and the basis of locomotion on land by animals all have their roots in the spine. Showing the impact of modularity theory upon our understanding of the emergence of these features will be a major focus of the latter portion of this book.

From Proteus to Modularity Theory

In this book, I will trace the origins of this modularity concept in the late 18th century, its development and later abandonment in the 19th century, and its gradual restoration and expansion during the 20th and 21st centuries. Using biological examples, I will show that a Modularity Theory can be outlined that is far more effective and complete than either a purely adaptationist (Darwinian) or purely structuralist (evo-devo) approach to the explanation of similarity among animals.

I will show how the perspective of Modularity Theory can revolutionize our view of human origins. I will demonstrate how it can change our understanding of processes such as the "Cambrian explosion," in which all 35 existing animal phyla rapidly emerged 500 million years ago, many from a common "urbilaterian" ancestor. I will recount its effect on our understanding of the rapid origin of the modern mammalian orders 70 million years ago. Finally, I will describe how it can explain functional transformations such as the emergence of bipedal, warm-blooded, carnivorous dinosaurs, as well as the origin of the body plan of the vertebrate phylum from a novel embryological pathway that can be characterized as an "inverted insect" pattern. This theory pulls together in a coherent form some of the most astonishing biological discoveries of the past 20 years and applies them, along with 200 years of biological research, to reinterpret the biological world. As these discoveries have accumulated in the last two decades, the most amazing findings have been lodged uncomfortably in the existing edifice of classical evolutionary biology. However, it is clearly time for a major theoretical change.

The best example of the limitations of the existing paradigm of the Darwinian Modern Synthesis in Evolutionary Theory arises in the consideration of Geoffroy's explanation of commonality between vertebrates and arthropods—now proven to be correct. Vertebrates have their nervous system in back (facing the sun) and their digestive system in front (facing the ground); insects have the opposite configuration. In the embryo, vertebrates form the anus from the embryonic structure called the *blastopore* (the initial "inpouching" of the hollow sphere of cells that appears very early in embryological development), whereas insects form the mouth from the blastopore. Geoffroy suggested that insects could be grouped with vertebrates if we just allowed for a change in which the body axes simply got flipped. Everyone laughed at the prospect of this idea until the specific single gene alteration responsible for exactly this change was discovered in the 1990s.

This is one of the most critical events in the history of animals. It is an event that took place more than 500 million years ago, yet it can be experimentally verified in the laboratory today. However, the verification of the event does not prove speciation, and it does not prove natural selection. It simply proves that the fundamental developmental plan that makes an insect can be changed *through a single genetic event* into the fundamental developmental plan that makes a vertebrate. Many other changes are required to actually produce a real vertebrate animal; however, the one gene change causing the inversion event transforms the entire organization of the animal.

Aside from the historical and biological drama inherent in this discovery, it poses a huge problem for classical Darwinian Evolutionary Theory. For the most part, Darwinian evolutionary change occurs more or less gradually among populations that share gradually changing mixes of genes until a *speciation event* (a genetic alteration that permanently splits the group into two separate species whose members are unable to successfully interbreed) separates the two populations. Natural selection affects the frequency of occurrence of gene versions, or *alleles*, in the parent population and then in the daughter populations. It's obvious that the flip of the dorso-ventral body axis happened in one single, sudden event. Aside from the fact that the offspring survived and were able to breed, the action of the environment that changes the genetic composition of species populations (natural selection) appears to play no role in the original innovation that leads to the vertebrate animals.

It is widely accepted that the environment plays only a very limited role in the actual creation of new genetic variants. Variability arises from random mutations, and natural selection then acts on descendant members the species. The type of qualitative alteration that results from a major body plan change—such as that which occurred at the moment of origin of the vertebrates—so totally alters the descendants that they are no longer subject to the same framework of adaptive selection as members of the parent species.

There have been numerous suggestions along these lines with regard to Evolutionary Theory over the past few decades. What has changed is the degree to which we can show that animal body plans can change abruptly. The concept here has been presented at two different levels by Stephen Gould: one for large patterns affecting large groups of species (macroevolution), and one for individual characters in a species (microevolution).

In the early 1970s, Gould drew much attention with his concept of brief bursts of generation of many new types of creatures followed by long periods of slow, stable evolution—a concept called *punctuated equilibria*. The major objection to Gould's idea—and an objection that eventually eliminated any direct impact on core Evolutionary Theory—was that the rapid bursts of change he was describing might not really have been all that rapid. Instead of seeing species change shape gradually over millions of years, he showed abrupt changes in the fossil record of some forms. However, there was no escaping the fact that an "abrupt" change might still involve thousands of years, maybe a million. In the end, it was difficult to see how this differed from the gradual change of Darwin's typical model.

The second and more easily understood—and more humorous—example from Gould on this subject applies to morphologic change for a single character. He was addressing the issue of how natural selection causes a new type of anatomical character to develop. For instance, the classical view suggests that a 2 percent gain in running speed might result in a 2 percent gain in relative success under natural selection. In this fashion, natural selection, across 50 generations, will cause a 100 percent increase in running speed. He then focuses on the dung beetle, which evades predation by resembling, well, dung. This is so potential predators will ignore it. Asking how this trait could evolve gradually in steps, Gould writes: "Can there be any advantage in looking 5 percent like a turd?"

What is different now is the hard genetic evidence for a major transforming change of the body plan happening abruptly in a single generation. This abrupt time frame is not a matter of the pace of change of large assemblages of species, and it is not a matter of speculation about how many generations it takes for an individual anatomical character to appear. Rather, it is a matter of asking how to deal theoretically with an event in which virtually every structure in the body becomes greatly transformed all at once in a true instant—literally less than a second for the single mutation to occur. And it is no less important an event than the initiation of the entire vertebrate lineage.

It is tempting to say that the existing Modern Synthesis theory of evolution works almost all the time, so why worry about altering it if it doesn't do so well for one or two very unusual events? Instead, I consider the issue to be similar to the relationship between the physics of Isaac Newton and the physics of Albert Einstein. Newton's simple laws of motion, mass, inertia, gravitation, and so on had seemed definitive and immutable. However, Einstein showed that Newton's laws and equations were only approximations that appeared to be correct when examined at the common slow speeds in which the contribution from relativity is so small as to be negligible. The actual equation for mass and energy have to include a relativistic factor to be truly accurate, even though this additional factor has very little effect unless the objects are moving at a speed very close to the speed of light.

Similarly, the Modern Synthesis correctly describes most evolutionary change because most of what takes place does appear to be gradual, based on natural selection acting on a vast pool of random variation, and subject to the statistical rules of population genetics. Nonetheless, that paradigm cannot explain some of the most important transformations in biology without being expanded to accommodate the view from Modularity Theory and the morphological genetics of embryological development.

What is evolution when it occurs without the benefit of population genetic effects? The key point here is that an expanded Evolutionary Theory now must be elaborated so that it can allow for change under the paradigms of population genetics or morphological genetics. When Darwin wrote his ideas in the 1860s, genetics did not play any role. Therefore, it is the synthesis of Darwin's ideas with population genetics that is most affected by this expansion of Evolutionary Theory.

Darwinian Evolutionary Theory is overwhelmingly valid and correct, but its incorporation into Modularity Theory shows that Darwinian evolution is *necessary but insufficient* to explain similarity among animals. Similarly, the evo-devo program is also correct and necessary without being sufficient. The generation of major changes in body organization or body plan clearly does play a major role in explaining animal diversity. Arguments about the role of body plans in evolutionary change have gone on for decades, but what has changed is that we now have a far more precise understanding of how these changes take place.

Further, one more very fundamental thing has happened: The "rules" that seem to limit the types of morphological (anatomical) change that can occur are starting to be understood. When Goethe or Geoffroy imagined some sort of fundamental unity of plan, the idea seemed to be metaphysical. Therefore, biologists such as Darwin felt free to dismiss and ignore the details of the subject. Now it is clear that these constraints are not only real, but may be more readily subject to direct laboratory investigation than any other underlying aspect of evolution.

As I will demonstrate in this book, some homeotic changes are closely engaged in adaptation, whereas some appear to be almost completely disengaged. The biological significance of these changes can be assessed by their intrinsic mechanism—their origination in changes of morphologic gene function. Whether they affect structure, developmental mechanisms, or environmental fitness is a secondary matter. Their realm of effect is interesting and worth studying, but it is not the source of their biological importance.

Change in homeobox genes is important because nature effectively tells us that it is important. Some genes change slowly and some change rapidly. What we know about many *Hox* genes is that they virtually do not change at all over vast stretches of time. We don't really even have a good explanation of how these genes can be so resistant to change—what about the gamma rays, replication errors, and so on? The only well-understood mechanism for this effect would be the immediate death of any embryo in which any change occurs. In fact, it is not only the genes themselves but their order in gene complexes and the nature of their controlling factors (called *cis-acting* enhancers and repressors) that remain remarkably stable.

It is high time to reassess the relationship of the major historical, philosophical, and biological contributions to understanding the History of Life on the planet. In doing so, new data from molecular biology, anatomy, and paleontology, as well as the contributions from new and emerging theories of underlying biological processes, can greatly expand and strengthen the explanatory power of Evolutionary Theory. The result is a new understanding of who we are, how we came to be, and the true nature of the world around us.

Goethe's Poetry of the Spine

Introduction

There is a spelling or a series of letters or words meted out in DNA that describes the origin of the spine. This is the spelling of the *Hox* gene family. These genes, with their shared homeobox DNA sequence, provide the instructions for organizing the segmental base that is the core anatomical element upon which the vast majority of animal species are constructed.

The *Hox* complex is the modern-day physical manifestation of the magnificent and lyrical idea that burst into great poetic mind of Johann Wolfgang von Goethe one blustery morning in 1790. It was an idea so powerful and so beautiful that it swept along cynics and believers, scientists and synods. It was an explanation of how the animals of the earth had been created, and it was so effective, scientific, and comprehensive that virtually every biologist in the three generations before Darwin accepted it as fact, and virtually every theologian accepted it as miraculous proof of how God's hands had shaped the creatures of the world.

Goethe had been fascinated by the seminal scientific event of his century—namely, Lavoisier's demonstration that all materials could be reduced to repeating small elements. He became convinced that the similarities among animals must reflect some underlying elemental structure. Then, during a morning walk in at the Lido beach in Venice in 1790, a friend tossed a weathered sheep skull into his hands. The sheep skull was falling apart in such a fashion that Goethe saw in its pieces a similarity to the vertebrae of the creature's spine. This became the basis of his theory. If the skull of a sheep could be made from modified vertebrae, then perhaps any structure in the body could be generated in similar fashion. He imagined the vertebra as a fundamental animal segment reduced to its purest essence. The segments existed as a string of radially symmetric objects, not unlike the crystals being discovered in what was then the newly born science of chemistry.

43

A Second Version of Goethe's Theory

As it turns out, Goethe really made two passes at his project of understanding the fundamental composition of living things. This is not surprising if you follow his works. Students of Goethe's life will know that this was a man who was constantly reinventing himself and revising his own ideas. Despite tremendous success as a writer, he became a government bureaucrat, Chief of the Department of Mines, and, just as suddenly, after 10 years in government, slipped away to Rome under an assumed name to study art, to write, and to contemplate the basis of nature. Many of his written works were scattered in notes and phrases and left unpublished in stacks for decades at a time. Nonetheless, he produced numerous literary works so greatly respected that even today, centuries later, there are still tens of thousands of Germans who have read nearly every word he ever wrote.

Goethe's works are more difficult to follow and appreciate in their entirety for non-Germans, as existing translations are only variably adequate. Perhaps this can explain one of the most stunning misunderstandings in the literature of biology and of the history of science. Goethe is generally cited as the source of the concept of the pure "typus," a term later expressed by others as the "archetype." This is the form of an elemental, essential creature that could be imagined as a source design or template for all other animals. It is true that this concept figured prominently in his earlier works and was championed by later authorities who credited his creative role in the development of the idea. However, as I will soon demonstrate, Goethe had a better idea later, and this second idea has been missed to a great extent. This is unfortunate, because it is this second idea that lends the greatest predictive power to his biological concepts.

The second idea held that all complex life—plant or animal—was assembled from modular units. The idea of the same object appearing in an essentially infinite variety of forms and being linked together by a common essential design was based on Proteus, the Greek god who was capable of assuming many forms. This Proteus concept was not some sort of pure abstract philosophical idea; rather, it was a way of describing the modular nature of life. Complex plants and animals could be assembled from repeating combinations of fundamental elements. For plants, his "monomer" was represented by the leaf. For animals, it was represented by the vertebrae, or the segments that line up along the main longitudinal body axis. A partially completed drawing in which Goethe depicts the leaf as the plant monomer, and various versions of insect segments for the animal monomer, is almost chilling in its anticipation of modern understanding of the issue (see Figure 2-1 on the opposite page).

Each of the animal segments would be radially symmetric, with four extensions pointing away from the center of a circle, similar to the principal directions of north, south, east, and west on a compass. The number of segments and the shapes of the extensions could be varied. From this fundamental set of building elements, the shape of virtually any animal—whether mammal, reptile, insect, or sea creature—could be generated.

The first idea—the archetype—did not really employ the reductionist approach (understanding something by evaluating its component parts) that led Lavoisier to discover the theoretical basis of chemistry. Rather, it captured Plato's ancient classic concept of the *eidos*—the fundamental idea that, through variations of its details, could serve as the basis for an intellectual approach to understanding any field of knowledge. It was his later idea of the protean monomers, however, that bears the greatest relevance for modern biological science.

Impact of an Idea

Goethe is a remarkable human figure. He was the greatest poet and playwright of his time, and among the greatest novelists who ever existed. What some people do not realize is that he was also the author of some of the most im-

Figure 2-1: Goethe's sketch of modular units in plants and insects. This partially completed drawing by Goethe included in his 1790 *Zur Morphologie* shows that he appeared to be looking for a single approach to a fundamental monomeric module in both plants and animals. Beyond the vertebral theory for which he is famous, this drawing shows he was trying to develop a concept that incorporated insect segments as well (Goethe 1790).

portant scientific works of his time: one in the physics of color, one in botany, another explaining the nature of clouds, yet another in geology, and finally a collection of important works in comparative anatomy.

His career was unusual in many ways. *Faust*, his most famous play, is the story of a man who sells his soul to the devil in exchange for knowledge and power. He wrote the first draft when he was 25 years old in 1774, but the first performance of *Faust Part One* did not take place in Germany until 1829, when he was 80 years old. Then, in the next three years, and ending with this death in 1832, he completed the even more important *Faust Part Two*.

Although Goethe traces his ideas about animal form to insights he experienced in 1790, he did not publish his zoological work until 30 years later, in 1820. The typology of his first theory of organisms, followed by the monomerism of his later works, is similar to the trajectory of Goethe's two works on Faust. A major and widely recognized first work, *Faust Part One*, is then followed decades later by a second and very different work, *Faust Part Two*, which requires at least equal attention.

How We Understand Goethe's Work

In modern Western science, it is almost a given that scientists will examine and consider the information in a scientific publication without assessing who the writer is, their mood, their background, their intentions, or their state of mind. Essentially, scientific publication is a very formal, structured type of communication. The writer or writers typically first explain the scientific background for their work. This generally includes a series of references to previously published findings by other scientists working in closely related areas. The writers then identify a focused remaining problem defined by what has gone before, and propose a theory that takes a small step forward—generally predicting what everyone in the field believes is most likely to come next. Following this is a "methods" section, in which the writers detail precisely how the work was done so that any other scientist in the field could completely reproduce the work. Next, the specific results of the tests are reported. Finally, the author or authors explain their view of how the findings support or negate the hypothesis at the beginning of the paper. Other scientists will do similar work, and a consensus will eventually develop as that scientific field moves forward.

In English literature, a similar situation often prevails. Although there is a good deal more interest in biography about important writers than there is about their scientific counterparts, in the end, the interest lies almost exclusively in the text itself. Shakespeare is as good an example as any: We know a great deal about what he wrote, about his writing styles and methods of expression, but we know very little about the fine details of his life as each work was being written, edited, put into production, criticized, and so forth.

Johann Wolfgang von Goethe is an entirely different kind of phenomenon from both a literary and scientific point of view. As a scientist, Goethe is intensely autobiographical. He writes constantly about himself—what he is doing from day to day, whom he has talked to, what he thinks of his critics and contemporaries, what he intended to accomplish on a given day, and what distractions intervened. He shares letters to friends and mixes them in with letters to his old professors, letters to his employers, comments about relationships, women he is interested in, art he is seeing, clever new ideas for a play he is working on, and so on. He tells us how he loves to dramatize whatever he thinks about, and he sets down on paper the dramatic shifts in his own thinking—demonstrating a great enthusiasm for a particular idea or writer, coming to a decision that the work or writer is really not so great, and finally rewriting his own work that had once been so influenced by the now-reviled source.

Again, a huge obstacle in understanding Goethe for the non-German speaker is the language—in translation, all of the lyrical word play that appears in his work is effectively lost. Additionally, most writing about Goethe comes from a literary world that probably knows little about biological archetypal theory or modern modularity theory, and certainly hardly anything at all about vertebrae or the embryological development of the skull. Finally, scientists, particularly those in the English-speaking world, tend to be baffled by his writing.

In trying to understand the significance of what Goethe wrote almost 200 years ago about the similarities between animals, it is important to reflect on the fact that his most formal work in biology was his paper on the human intermaxillary bone (see Figure 2-2). This work is sometimes credited as a key document in biology because it uses observed anatomical evidence to prove that humans are indeed part of the continuum of animal life. However, what he discovered was already published by other more traditional scientists of his time (particularly Felix Vicq D'Azyr; see Chapter 3), and, despite his literary fame and government position, no scientific or academic press would publish his work for almost 30 years.

Goethe's formal scientific anatomic work in the intermaxillary paper seems to have been a bit of an aberration for him, and he doesn't bother with a traditional scientific approach in biology after this endeavor. Fortunately, what we do have is a remarkably detailed, day-by-day autobiographical accounting of a few critical months in Goethe's life in the spring and summer of 1787. For the English-speaking reader, these thoughts have been published in a translation that is the result of a collaboration between an experienced Goethe translator, Elizabeth Mayer, and the accomplished English poet and literary figure, W.H. Auden. The text of *Italian Journey* and the introduction by the translators goes a long way toward helping the English-speaking reader understand just what it was that Goethe was up to. However, a new translation of sections of *Italian Journey* was completed for this book by Elizabeth Jackson, with review by Daniel Schaeffer. This entirely new rendering reveals important biological features of the text that were hidden in Mayer and Auden's version.

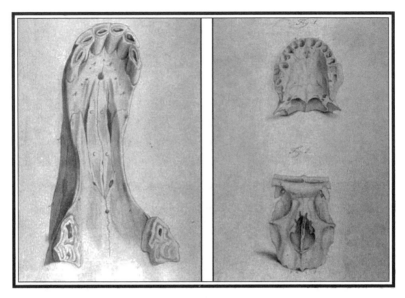

Figure 2-2: Goethe's report on the intermaxillary bone in humans.
This formal illustration from Goethe's publication shows his discovery of a bone
that anatomists of his time had claimed was missing from humans. This was used
to support the argument of continuity of human anatomy with the rest of the biotic
world (*left*, horse; *right*, human).

A Meteoric Rise to Fame

Most of the information in *Italian Journey* is concerned with Goethe's
ideas in botany, but a great deal is revealed about the project and how he went
about it, which is relevant to understanding his accomplishment in biology in
general. In brief summary, Goethe achieved spectacular literary fame as a
young man first with a play—*Götz von Berlichingen*, about a 16th-century
military adventurer—and then with an enormously successful novel—*The
Sorrows of Young Werther*, a tragic tale of love unfulfilled—published when he
was 26. The novel was, in its day, shocking, trendy, and provocative. However,
it made him wealthy and one of the great celebrities of his time. It is said that
thousands of young Europeans committed suicide after reading his novel as
they identified with the agony of its lovelorn protagonist.

As Goethe approached 30, however, he seemed to be unable to figure out
how to match his youthful accomplishment. Without any warning or plan-
ning, after a chance meeting with a member of German royalty, he accepted a
spur-of-the-moment offer to move to Weimar, and to take a government job
based on his obvious genius. Part of this job involved making himself available
to entertain and enlighten the court. The young Weimar prince who made

the invitation thought the popular Goethe would be interesting to have around in the provincial capital. Goethe apparently thought it would be a good idea at the time to take a long-term, secure government job.

In Weimar, along with his various government duties (including finance, war, and mines), he studied anatomy, dabbled in science, and worked on several major literary works. All of this proceeded over the next 10 years, as he developed a pattern of steady, if not brilliant, output in a wide array of fields while living the life of a pampered celebrity in a European capital. Then, one day in late August of 1786—apparently just after learning of Lavoisier's great accomplishments in chemistry—Goethe disappeared from the Weimar court. He assumed a secret identity and took a coach to Rome. Six days later, he wrote saying he would be gone for a while.

The Emergence of Biological Theory in Rome

During the year in Rome that followed his sudden departure from Weimar, Goethe immersed himself in art history, architecture, the classics, the modern art of his day, and the heady vigorous life of a single, well-off foreigner in the Italian capital. As he was relatively unknown in Italy, he abandoned the celebrity life, spending his time instead primarily with some expatriate German artists.

The Primal Plant

On the 27th of September in 1786—just a month after disappearing from the Weimar court—Goethe writes:

It is exhilarating and educational, to roam amongst vegetation that is unknown to us. With familiar plants, as well as with other things commonly known to us, we, in the end, give no thought to them, and what is contemplation without thinking? Here, where I am faced with *this advancing diversity*, that thought becomes again, ever more alive, that *all plant forms in existence derived from one*. With this theory, only then would it become possible to determine sexes and kinds precisely, which to me, thus far, happens very arbitrarily. Upon this point, I am stuck in my botanical philosophy, and I do not see yet, how I am going to untangle myself. The profundity and repercussion of this business seem to me equally weighted. (emphasis added)

In the paragraph before this, Goethe is criticizing the architectural design of an Italian medical school. In the paragraph after, he is talking about stalls in an annual fair in Padua. Nonetheless, the reader of this book will recognize the electrifying nature of his assertion. You can see exactly what he is seeking. His vertebral theory of the skull—which he first mentioned to others four years later—is not a humble scientific discussion of sheep skull embryology: The man wants nothing less than a unifying explanation of the underlying basis of the similarities among organisms. He is setting out to understand and explain something extraordinary and fundamental that will transform our understanding of nature.

Five months later, on February 19, he is in a foul mood walking through the Villa Medici. It is in the evening, and he is wandering in a garden. He spies a telescope and examines the edge of the dark side of the moon's sphere, and then looks away from the telescope at some oak trees:

> Today was a day, spent painfully, amongst fools. I am on the way, toward gloriously new discoveries; how nature, with such enormous magnificence, out of that which looks like nothing, develops the most diverse from the simplest.

The new translation uses "diverse" instead of Auden's "complex," thus emphasizing the fact that Goethe had seen past the idea of a "scala natura" to envision progressive diversity. Here again, the scientific project is clear: He is looking for the fundamental building blocks.

Essentiality of Defining Form

Then on April 17, he decides to go to the Public Gardens in Palermo to reflect on some new ideas for a play. However, a plant catches his eye:

> In the face of so many new and renewed formations, that self-doubting question again came to mind, if I could not discover the Urpflanze [archetypal plant] in this prolific profusion, could it exist? There must be such a one! How else would I recognise, that this, or any other formation was a plant, if they weren't all created using the same original pattern?

Here we have the core of the idea. This is clearly the concept behind his animal archetype and the initial motive behind the project. He continues in the following paragraph:

> I've tried to examine, the ways in which these figures differ from each other. But found more similarities than differences; and when I tried to apply my botanical terminology, it could be used, but didn't prevail. It unsettled me, and did not push me to go further, as sometimes a challenge can. My good poetic intention was unbalanced, the garden of Alcinous had dissappeared, replaced by a world garden. Why are we, so easily disturbed by what is new, why are we so perturbed by obstacles that we cannot overcome!

On the May 17 he writes to his old professor, Johann Gottfried Herder:

> Furthermore, I must confide in you that I am very close to discovering the secret of how plants multiply and are organised, and that it is the

easiest thing that can be conceived. One can make the most beautiful observations beneath this sky. I have clearly and without doubt identified the principle point, where the germ lies; I see everything else as a whole, and there remain only a few points to be clarified. The Urpflanze [archetypal plant] will be the most wondrous creation, for which Nature shall envy me. With this model, and the key to it, one can immediately invent an infinitude of plants which must be consistent; that is, they, if they do not already exist, could exist and are no pictorial or poetic shadows and illusions, but possess an inner truth and imperative. The same law can be used for all other living things.

Here, he not only provides further details of his developing ideas about plants, but he also explicitly states that this conceptual approach to botanical biology (Primal Plant, rules for construction, and so on) will apply to all other living organisms—that is, the animals.

A Conceptual Break of 30 Years in One Month

The first two books of *Italian Journey* were assembled for publication in the late 1780s. However, the third book, which describes events beginning in July of 1787, was not prepared for publication until 30 years later. Mayer and Auden complain that Goethe has done virtually no editing before he published. However, as a scientist reading on the subject of the Primal Plant, I can see that he may have weathered considerable criticism in the subsequent decades. From his other autobiographical writings, we know that Goethe loves to ridicule his critics in writing. Nonetheless, the text of the events in July differs considerably from his writing about those in May, primarily in that it describes details that are more typical of a scientific paper.

A Flash of Inspiration: The Birth of a Modular Theory

On July 31, Goethe writes:

While walking in the Public Gardens of Palermo, it dawned on me that hidden in that organ of the plant which we normally describe as the leaf, inherently lies the true Proteus, which can conceal and reveal itself in all manners of form. Anyone who has experienced within himself what a complex thought conveys, whether it comes from within or imparted and instilled by others, will bear witness to how passionately our spirits are moved by it; and how enthused we feel when we anticipate the entirety of that which develops bit by bit, that to which this development leads. Bearing this in mind, I can be forgiven for being captivated and driven by such an awareness as if by a passion, and being compelled to occupy myself with it—not exclusively, but for the rest of my life. But as much as this affinity had taken hold of me within, there was no question of my thinking of taking up formal studies after my return to Rome; poetry, art and antiquity all effectively challenged me completely, and in my whole life I have never easily spent more wearisome and laboriously busy days. It may sound all too naive to experts in the field when I relate how every day, in every garden, on walks and small pleasure trips, I would take possession of the plants I noticed beside me. It was especially important for me to observe how, as the plants went to seed, some of them saw the light of day. So I paid much attention to the germination of the cactus opuntia, which is formless as it grows, and saw with pleasure that it revealed itself, perfectly innocently and dicotyledonously, bearing two tender leaves; but thereafter, as it continued to grow, its future formlessness developed.

He then follows with a discussion of the details of growth from budding and a description of flowering in the carnation plant:

> Seeing before me no means to preserve this wondrous design,
> I took it upon myself to draw it exactly, and thus I gained an
> ever greater insight into the basic principle of this metamorphosis.

Thus, in the first two books, his descriptions of the Primal Plant theory are general, personal inspirations gained from observance as he wanders through a garden at night. In the third book, he is developing his ideas by collecting many specimens, describing his painstaking observation of actual germination of plant seeds, and completing drawings of key observations for subsequent study and possible distribution. The descriptions of his ideas in the first two books are those of a poet exploring nature through philosophy, whereas in the third book—set just two months later, but written 30 years later—he is at pains to depict himself as being vigorously and fully engaged in the objective and descriptive scientific methods of his day.

However, when you compare his thoughts of May of 1787 to those of July, there is a far more spectacular intellectual advance and transformation that takes place in Goethe's theoretical framework. Although they are set in *Italian Journey* as if they were written just two months apart, the final writing of the two sections is actually separated by the elapse of 30 years of his life. In the first two books, Goethe's vision falls into the classic paradigm of typology—he is seeking a Platonic vision of the perfect original plant. This is how his ideas are almost always portrayed in the history of science. In fact, when he returns to botany in the third book, starting immediately with his thoughts on July 31 of 1787, he has abandoned typology in favor of a far more advanced and far more profoundly useful concept from the point of view of modern science. He now is seeing Proteus in the leaf itself. It is the idea of a modular monomer that can be duplicated with modification to produce any plant. This is entirely different from an archetype, which is the primal animal or plant. Instead, it is the concept of the key role of repeating elements in the modular assembly of any biological entity.

Finally, on August 28, he writes again to Professor Herder:

Within the field of natural history, I am bringing things with me, which you are not expecting. I believe, the how of the Organism is not far from my reach. You should view these manifestations—not fulgurations—of our God with elation, and advise me, who in ancient or present times has made equivalent discoveries, had the same thoughts; including, those considerations that may deviate slightly from my own.

The Vertebral Theory and the Priority Dispute

Goethe made a brief second trip to Italy in March of 1790, traveling primarily in northern Italy. It is during this journey that he reports an event that is analogous to his September 1786 insight about the Primal Plant, his May 1787 revelation that animals could be explained in similar fashion, and his July 1787 epiphany in seeing the leaf as the Protean assembly element in plants. On a Venice beach, handling a weathered sheep skull, he experiences a moment of inspiration in which he realizes that the vertebra can serve as a guide to the fundamental monomer in animals, just as the leaf serves this function in plants.

Goethe insists that he started writing up his theory later that year (although, as with many of his other works, he did not manage to actually publish his vertebral theory of the skull until more than 25 years later). His report on the leaf and some more general thoughts on the application of an archetype idea to animals did get published in the 1790s. It may be that he was still struggling with his concept of "archetype versus monomer," or it may just be a result of the usual vast array of distractions he encountered. In this case, however, his delay in publication resulted in a very painful and public controversy that troubled him greatly in his later years, and that has continued to lead to doubt and uncertainty about his version of events.

Oken's Position

The competing claim for the vertebral theory of the skull from Lorenz Oken has to be seen in this context. Oken was born in 1779, and so would have been 11 years old when Goethe tells us he had his 1790 insight about the vertebral makeup of the skull. In *Italian Journey*, Goethe tells us that he had all this in mind (see his comments of May 17, 1787) when Oken was only 8 years old.

Oken published his vertebral theory in 1808, but Goethe did not publish *Zur Morphologie* until 1817—the same year he finally published *Italian Journey*. Oken aggressively and angrily expressed the belief that his work had priority. In 1847, writing in a preface to third edition of his works, Oken states: "I have shown that the head is none other than a vertebral column.... This doctrine was at first scoffed at and repulsed; finally, when it began to force its way, several barefaced persons came forward, who would have made out if they could, that the discovery was achieved long ago." More to the point, Goethe actually invited the 28-year-old Oken to come to Jena to take up a professorship. At his inaugural speech upon arriving in Jena in 1807, and with Goethe in the audience, Oken first presented his vertebral theory.

Over the past 200 years, this priority dispute between Oken and Goethe has been viewed as an entertaining historical footnote about contention over an idea that itself has long been discredited (Richards 2002). However, a case can certainly be made that this seemingly tangential debate actually relates to one of the most central defining insights into biology of the past millennium— and perhaps one of the most far-reaching scientific insights of all time.

Although there are now numerous books and articles about this priority dispute, the issue has remained unsettled. The various accounts differ for several reasons. Some are written by biologists who don't realize that they must attend to the philosophical and theoretical aspects. Some are written by historians who don't understand the biology. There are accounts written in German in the 1800s with excellent access to a wide array of relevant primary documents, and there are accounts written in the past five years that rely on relatively recent reviews of limited sets of documents. I believe that with a careful review of the underlying historical, philosophical, and biological issues, I can lay this issue to rest. I will state my conclusion at the outset: Goethe deserves the credit.

Oken is correct that Goethe had never published his views about vertebrae until after 1807. However, there are three major reasons why Oken's claim to priority cannot be accepted: firstly, the idea itself was not entirely novel; secondly, what Goethe had to say about it was recorded in a variety of personal letters written before 1807; and thirdly, the theoretical significance of what Goethe had to say goes far beyond what Oken contemplated. Comparative anatomists who are reading this well know that it doesn't take much imagination to appreciate the relationship between the occiput of the skull and highest vertebra. In fact, a number of German anatomists without any special claims to intuition had pointed out the fairly obvious construction of the skull base out of three vertebrae. These ideas were available to Oken and to Goethe at about the same time. Oken studied anatomy at the Universities of Wurzburg and Gottingen between 1801 and 1805, and Goethe learned anatomy in Jena in 1780. Works by Johann Autenreith, Carl Friederich Kielmeyer (who trained Cuvier in anatomy), and Johann Peter Frank (a professor at Gottingen), among others, covered the issue without a great deal of fanfare. Goethe and Oken both proposed that the entire skull could be made up from vertebrae rather than just the skull base, but this was a matter of additional evidence and interpretation extending an idea that was not entirely novel. Nonetheless, Goethe felt that the sheep skull he saw in Lido confirmed the potential for deriving the entire bony skull from modified vertebrae. He immediately communicated his observation in letters.

In the United States Patent Office until quite recently, personal notes shown to one other person could be used to establish priority for an invention. The reason for this protective measure was the assumption that another person working in the same field would readily learn of the idea and try to file first. In science in general, however, when an important observation is made, sending a report about it to a few friends and then leaving the idea alone for 30 years might not be accepted as evidence of priority. This is because there is a presumption that an idea by itself is not substantive; only in the context of a fully worked, formal publication that reveals method and data and is reviewed by peers, is an idea actually made manifest and validated. Goethe's letters, however, differ in a fundamental way from a letter written by anyone else. At the time, he was the foremost celebrity in the world, and his every written word was carefully attended to.

For decade after decade, he scribbled comments, phrases, paragraphs, and statements on pieces of paper that were later collected, assembled, and published. He leapt out bed in the middle of the night to jot down sudden insights. He wrote the same thing more than once because he did not always remember what he had already written. It is fascinating to consider that his personal secretary, Johann Eckermann, who assembled a mass of this material into several works, wrote a summary of his conversations with Goethe (Goethe et al. 1852). Nietzsche himself called this compilation the greatest work ever written in the German language. The point here is that every note scribbled by Goethe, and particularly every letter sent by him, was a matter of wide interest in his own time immediately after he wrote it.

Resolution of the Priority Dispute

The best confirmation that Goethe communicated his idea to other scientists when he claims he did comes from the work of Rudolf Steiner, who spent eight years in Weimar between 1888 and 1896, editing the complete body of Goethe's scientific works (Steiner et al.). Steiner reports that on April 30, 1790, Goethe showed the actual skull to Charlotte von Kalb in Weimar and explained his interpretation. Von Kalb was a close friend of Goethe's companion, the poet Friedrich von Schiller, and was well known in Weimar literary circles. According to Steiner, Goethe asked her to pass the news along to his old professor Johann Herder, with whom he had been corresponding during the initial development of his ideas in Italy: "Tell Herder that I have come a whole formula closer to the animal form and its various transformations—and that through the strangest accident," he wrote. Apparently, Goethe then wrote to Herder's wife, Karoline, on May 4, 1790, again reporting that he found a skull allowing him "to take a great step in explaining the formation of animals" (Richards, p. 420).

Aside from the poorly documented evidence of the events of 1790, Oken's claims that Goethe made the whole thing up after hearing his 1807 lecture are undermined by some correspondence dating from late 1806 to February of 1807 between Goethe and one of his associates, a botanist and physician named Friedrich Siegmund Voigt. This correspondence, initiated by Goethe, is explicitly about the vertebral composition of the skull. It clearly predates Oken's arrival in Jena and his first public report of his ideas. Beyond that, Oken's claims to have thought of it earlier are no easier to support than Goethe's more believable claim.

It has been suggested that Oken must not have known what Goethe had said about vertebrae and the skull. For instance, it has been pointed out that it would have been sheer madness for Oken to present this as his own idea at his inaugural lecture in Jena—especially if he knew that Goethe himself was the source of the idea. He was a 28-year-old in desperate financial straits who had just been offered a job by the greatest man in Europe. To publicly plagiarize his benefactor immediately upon his arrival would have been bizarre at best, and career suicide at worst. Oken attributes his idea to the finding of a deer skull in the Hartz mountains in 1805. Perhaps he had heard the story of Goethe's unpublished discovery but was not aware that Goethe was the source. Of course, if Oken had waited a few months after his arrival in Jena before delivering the lecture, he might well have learned of Goethe's idea—but this is not how events unfolded. The general assumption is that there was no way Oken could have learned of the story, so he must have stumbled upon his own inspirational decomposing skull, thus unknowingly duplicating Goethe's experience.

Charlotte von Kalb may be the key to resolving this enigma. In 1793, she hired the young poet Friedrich Holderlin to serve as tutor for her child. We know that during his year in the post, Holderlin wrote to the philosopher Friedrich Schelling about Charlotte von Kalb, and about traveling with her and her son. Holderlin was also in communication with Schiller during this period, and Schiller would have confirmed von Kalb's story. Charlotte von

Kalb was an interesting figure. Unhappily married to a man devoted to his military career, she was widely known to be the inspiration for the female characters in some of Schiller's dramatic works. She was also a writer, and, after Schiller married, she was associated with various members of the intellectual circle in Wiemar.

Schelling was a prominent idealist and one of the fathers of *Naturphilosophie*, the genre of naturalist contemplation in which both Goethe and Oken were involved. In 1805, the time when Oken supposedly saw his deer skull, he was corresponding heavily with Schelling and including frequent pleas for money. Oken's financial crisis only lifted when Goethe invited him to Jena, whereupon he immediately presented his vertebral skull story upon arrival. Did an irritated and put-upon Schelling "set up" Oken?

It's possible that Charlotte von Kalb entertained her son's tutor with her story about Goethe's sheep skull. The tutor could have told the story to Schelling, who, upon learning that Oken had been hired by Goethe to teach in Jena, could have told Oken the skull story without explaining that Goethe was the source. He then could have recommended that Oken impress every-one at Jena by revealing this fascinating idea at his inaugural lecture. If this is true then the echoes of this practical joke have continued for 200 years.

A nagging question still remains: If the idea was so important, why didn't Goethe publish sooner? In part, Goethe's complaint about Oken is that the publication is too simplistic. Goethe implies that he was still working out all the details of the idea. However, as with Darwin, the context also needs to be seen. *The Origin of Species* was published nearly 30 years after Charles Darwin had reached his principle conclusions. The intervening years were spent in collecting additional data and fleshing out his ideas. (It could be pointed out that Darwin at least learned the lesson about priority of publication from Goethe's famous dispute with Oken. Darwin realized that he had to publish immediately when he received a letter proposing a theory of evolution from Alfred Russell Wallace, who was doing field work in Borneo). Moreover, there was the simple practical matter that, despite his fame in literary circles, Goethe was having a difficult time getting published in science in the 1790s. He was subject to several criticisms—he was a poet and dramatist trying to do science, his ideas did not agree with those of the leading biologists, and his methods did not display an adequate level of objectivity in the observational data. He was already locked in a bitter struggle to convince the scientific societies and journals to allow him to publish his now famous discovery of the human intermaxillary bone. (It was a critical point in support of the idea of unity in nature because it showed that humans had all of the same bones as other primates and were not uniquely designed, at least with regard to the exact number of bones in our bodies.)

Although Goethe is generally given credit for the date of the discovery in 1784, he was unable to get the finding published for more than 20 years. When he eventually tired of his efforts to get published in existing scientific journals, he began self-publishing. The leading biologists of his day had made a major point of the absence of the intermaxillary bone in humans. It was therefore highly offensive to them that that this poet had chosen this point to refute so publicly. They tried to talk him out of it; they blocked publication; and they launched attacks on him that he bitterly ridiculed in turn, just as he ridiculed his literary critics.

On the upside, Goethe was able to publish his theory of plants in 1790. He was also heavily involved in his theory of color. Indeed, his contribution to optics is interesting and often neglected by biologists trying to understand his theories in animal morphology. Although his writings are filled with reverential praise for Shakespeare, he goes on endlessly about Newton's failure to understand optics correctly. He identifies himself instead with DaVinci and Plato. Not only does he point out repeatedly how he beat Newton on this, but he is indignant and angry that academics refuse to accept his view. They appear to support Newton, even though the data show that Newton is wrong. Goethe attributes this to conservatism, to lack of respect for his science because he is a literary figure, and to simple self-preservation on the part of prominent academics who want to defend what they have always taught even if it means trampling upon truth.

He was writing and rewriting his great work, *Faust*. He was fighting his battle against Newtonian optics (Newton's ideas have won out in the end). He was publishing his plant theory. He was working with various translators to get his literature and scientific work published in English. And throughout all this he continued to expand his reflections on the archetype and later the modularity concept. However, as with much of his other scientific work, he did not publish. Nonetheless, the lack of publication did not mean that the whole world was unaware of what he was thinking.

The third point in favor of Goethe is the theoretical context. At least 95 percent of what Oken ever wrote or published in science was considered comical by many biologists of his day, and today that percentage is closer to 100. For instance, similar to Goethe, he published on the human intermaxillary bone, but said it was in fact identical to the thumb. The only memorable bit of real science from Oken was the vertebral theory of the skull; however, it is really Oken's claim for priority regarding this theory for which he is remembered. His theory was part of a bizarre and fantastical framework in which he pointed out oddly mystical similarities among biological structures. Although the idea about skull vertebrae is remembered and recited, virtually nothing that Oken wrote in science can be read usefully or even respected today.

In the case of Goethe, the opposite holds true. Virtually every word he ever wrote is still being read by tens of thousands of people. Republication and translation, and even third and fourth retranslations into various languages, are ongoing. This applies not only to his poetry and plays but also to his scientific writings, his aphorisms, and even to his random conversations. The reason for all the interest is clear when you start to read Goethe. He is no Shakespeare (at least when read in English translation), but the effect is similar. You always learn something of value and enjoy sharing in his insights. When read by a native German speaker, there is an entire additional layer of engaging, playful, and masterful use of language that draws readers in and creates the fascination and reverence for every single word that many describe.

Significance of Goethe's Biology

The significance of what Goethe had to say about vertebrae is not just a matter of historical reverence: It is entirely substantive. Goethe stood at a key point in the growth of Western thought. German philosophers of the late 18th century, such as Schelling, Herder, and Kant, were seeking underlying unity in nature. Goethe blended philosophy, an intensely romantic view of nature, and the discipline of an empirically grounded scientific naturalist to link philosophical expectations with actual biological phenomena. His theory of the leaf as the fundamental subunit of the plant is an enduring contribution. He identified the vertebra as the route to understanding the fundamental subunit in animals. His vertebral theory of the skull was not about explaining a comparison between the maxilla and a vertebral bone, as in Oken's case. Rather, Goethe was in the process of identifying where to look for a fundamental subunit out of which any animal could be assembled. That he chose to look at the vertebra was as prescient as it was correct.

3
The Remarkable Life and Indelible Impact of Étienne Geoffroy Saint-Hilaire

Introduction

It was September 2, 1792—year one of the French Revolutionary calendar. Across France, mobs had begun to carry out brutal executions of hundreds of imprisoned priests who had not given up their allegiance to Rome. Atop a ladder, in a small town north of Paris, a 21-year-old mineralogy student was risking his life to rescue a group of these priests. In this act and on this day, events that would lead to the development of a central element of modern biology were set in motion.

The young man on the ladder was Étienne Geoffroy Saint-Hilaire, and the imprisoned priests were associates of his favorite professor, René-Just Haüy (whose own release Geoffroy had won several days earlier). In a few months, the tables were turned: Haüy resumed his place as a leader of French science, and he rewarded his liberator. He brought the young Geoffroy to Paris and, just nine months later, helped make him the first professor of the new field of zoology in the first institution devoted to that field—the newly created National Museum of Natural History in Paris.

Geoffroy was to study the vertebrate animals, and a second professor, a botanist named Jean-Baptiste Lamarck, was recruited to study the invertebrates. A few months later, Geoffroy began recruiting another young man in his early 20s, Georges Cuvier, to work with him. Cuvier joined the museum in 1795, and the two became close friends, sharing a small apartment while launching the new museum and the new field in what they must have seen as a remarkable new world of opportunity and discovery. Four decades later, with Cuvier long since anointed as the world's leading biologist, the scientific and intellectual world would be electrified by the famous and bitter debate between these two former friends. The Great Academy Debate of 1830 set the terms for scientific research for the next two centuries, and laid some of the critical groundwork for the ultimate emergence of Darwinian Theory.

When news of the debate reached Germany, along with the news of the July 1830 revolution that ejected Charles X from the French throne, the great German poet Goethe knew that the debate was the more important of the two events. A friend of Goethe, Frederic Soret, reported:

> The news of the revolution of July, which had already commenced, reached Weimar today, and set everyone in a commotion. I went in the course of the afternoon to Goethe's. "Now," he exclaimed as I entered, "what do you think of this great event? The volcano has come to an eruption; everything is in flames, and we no longer have a transaction behind closed doors!" "A frightful story," I replied. "But what else could be expected under such notorious circumstances and with such a ministry, than that matters would end with the expulsion of the royal family?" "We do not appear to understand each other, my good friend," replied Goethe. "I am not speaking of those people at all, but of something entirely different. I am speaking of the contest, of the highest importance for science, between Cuvier and Geoffroy Saint-Hilaire, which has come to an open rupture in the Academy." (Appel, pg. 1)

Darwinian Theory is above all based on two ideas: that the world of nature changes across time and that it does so gradually under its own guiding hand. The biology of Geoffroy differs from the biology of Darwin. In essence, the most important idea that Geoffroy ever advanced—namely, the idea that the body form of the insect is abruptly transformed in a single event into the basic body form of the vertebrate animals—is not fully compatible with Darwinian Theory as it exists today. For 170 years after Geoffroy published his theory in 1822, this was treated as a wild and improbable rant from an otherwise important scientist—a view that was still held in an otherwise sympathetic work by Toby Appel published in 1987. Now, however, Geoffroy's idea is known to be correct; this change appears to have happened abruptly, leading to the origin of the vertebrate animals in no more than one or two generations in the ancient Cambrian sea (DeRobertis and Sasai 1996). However, as we will see, Darwinian Theory has not yet been adequately expanded to incorporate this spectacular biological event.

Science and Scientists

There is a "creation myth" for scientific progress that suggests that our growth of knowledge is steady, incremental, and predictable, no matter who the scientists are. This myth is the basis for the NIH (National Institutes of Health) and NSF (National Science Foundation) grant processes. Both direct the majority of their research money to well established, larger laboratories that are working on essentially similar problems, and to others in their fields using essentially similar methods but who demonstrate the best track record for solid work in the past.

However, there is an alternative view. What is it about Newton, Galileo, Copernicus, Darwin, or Einstein? Are they famous because of promotion, because of serial success, or because of political drama? In each case, we have a scientist who looks at a world of information available to numerous colleagues and has the insight and bravery to apply different methods and produce a dramatically different result from those of the other scientists around him. It helps that after the initial shock of hearing the new ideas, each has proven to be substantially correct in the claims he made—claims that, in each case, produced a scientific revolution.

The relationship between the story of a scientist's own life on the one hand, and the Story of Life told by that scientist on the other, is an important issue. The scientists whom we look to for evidence of our worldview tend to see the world in the same way they see their own work. Biologists embrace Darwin in part because the steady progress of species in his model of nature closely reflects the steady incremental advances by which they guide the plan of their own scientific work.

Étienne Geoffroy's story in science is utterly different and follows more closely the classic hero myth in Joseph Campbell's 1949 book, *The Hero with a Thousand Faces*, as typified by Odysseus or the more modern Luke Skywalker. A young man (Geoffroy) proves his extraordinary bravery and loyalty and is rewarded with scientific "kingship," becomes sworn to a noble cause, and goes on a great journey facing constant danger. Upon seeing the chambers of the gods, he experiences an epiphany, unlocks the secrets of the ages, and faces down his nation's greatest enemy on the field of battle to achieve immortal national honor—only to return home to find that his best friend has taken his throne. A final duel is fought to reclaim his kingship, his praises are sung by his descendants, and in the centuries to follow his great works are honored.

Commitment to an Idea—Unity of Plan

Geoffroy's dramatic career launch was only the beginning. Unprepared to teach a first course in zoology, he scrambled to educate himself and quickly stumbled onto a novel idea that became not only the theme of his career, but also, as we are only now coming to learn in the early 21st century, his greatest contribution to science. This was the idea of unity of design among the creatures of the earth.

During the year before his appointment as professor of zoology, Geoffroy's education included the teachings of his benefactor, Haüy, and of the chemist Antoine Lavoisier. Haüy had shown that a wide variety of crystal structures in the minerals of the earth could be constructed from a small number of fundamental crystal structures. And you'll remember that Lavoisier had shown that there were a number of basic chemical elements—oxygen, hydrogen, nitrogen, sulfur, and so on—out of which matter could be composed. In the biological world, however, there appeared to be thousands of independently created types of animals with no clear understanding of common features among them. Geoffroy intended to replicate what Hauy and Lavoisier had done in mineralogy and chemistry. He would look for a small number of fundamental structures out of which all the many thousands of animals had been constructed.

Geoffroy stumbled onto an approach to this problem in his first initial months as a zoologist as he began his research on one of his very first academic projects in zoology—a report about lemurs. Apparently, he found a paper written on the same subject a few years earlier by the famous French anatomist and physician, Felix Vicq d'Azyr. The idea of unity of design was outlined in the anatomist's works. Most unfortunately, Vicq d'Azyr's prominence had gained him a position as personal physician to Queen Marie Antoinette, and in the Paris of 1794, this was a serious liability. Geoffroy never had a chance to meet Vicq d'Azyr, as he had been imprisoned for his association with the royal family, and died of tuberculosis shortly thereafter at the age of 45.

The Journey

Another major event in Geoffroy's career occurred early in 1798 when an old acquaintance, a chemist named Claude Louis Berthollet, appeared at the museum. He met with Geoffroy and Cuvier to present the two young men with an unusual proposition. Napoleon was about to depart by sea on a secret expedition. Its purpose and destination could not be disclosed, but Napoleon wanted a team of prominent French scientists to accompany him. Cuvier politely declined, as he could readily see he had far too much serious work to do to even consider leaving his post for a few months. Geoffroy began excitedly packing his bags.

The secret mission proved to be Napoleon's invasion of Egypt. Militarily, Napoleon apparently wished to threaten the British position in India. Personally, he wanted his conquests to match those of Alexander the Great. There was also a tactical intention to break the waning strength of the Ottoman Empire. In any case, he chose to bring 167 scientists, writers, and artists with him to systematically study the entire land of Egypt.

A few months later Geoffroy would be climbing the pyramids of Giza as Napoleon's personal traveling companion. More importantly for science, in the three years between the initial glorious French landing on July 1, 1798, and the dramatic and famous departure from Alexandria under the guns of the British army in October of 1801, Geoffroy made many of the key discoveries and accomplished several of the critical theoretical insights that marked his spectacular career.

Discoveries in the Nile Delta

After Napoleon's quick conquest of Alexandria and Cairo, Geoffroy headed into the Nile Delta on his initial scientific expedition. From the start, military concerns were ever-present. Watching from a coastal hillside near the town of Rosetta, just one month after the triumphant landing, he was witness to the French disaster at the Battle of the Nile, in which British Rear-Admiral Horatio Nelson trapped and destroyed Napoleon's entire invasion fleet without losing a single ship. In a letter to his brother, Geoffroy described the battle in detail, including drawings that showed the horrible blunder of French Vice-Admiral Brueys that cost the lives of thousands of sailors and wrecked French military ambitions in the Orient. Apparently due to some incorrect data on local water depths, Brueys had managed to inadvertently render his powerful fleet defenseless. He moored his ships in a line along the shore, and Nelson realized there was space to sail between the French ships and the shore as well as on the seaward side. A double line of British vessels firing from both sides of each French vessel was overwhelming. It was an excellent example of how one simple bit of bad data from one man could have a lasting and significant impact on world history. Despite the naval victory, the British did not come ashore with troops, and so the scientific expedition proceeded.

The exploration of the Nile Delta took Geoffroy to Lake Manzaleh, where he discovered lung fish that could breathe out of water and walk across land with their fins. When news of the discovery of these fish reached Paris, Cuvier remarked that the entire expedition to Egypt would have been justified by the discovery of this one fish. Similar to the Rosetta Stone itself—found by one of the expedition's engineering officers in July of 1799, and which proved to be the key to deciphering Egyptian hieroglyphic writing—Geoffroy's discovery of the primitive fish *Polypterus* was the key to unraveling the apparent differences between the fish and the quadrupedal land vertebrates.

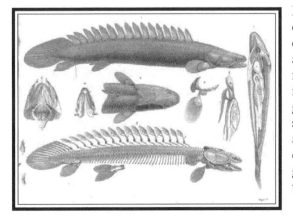

Figure 3-1: Geoffroy's figure of *Polypterus*. The discovery of this fish that could breathe air with its lungs and pull itself along on dry land with its four fins was one of the great events of 19th-century science, and probably played a greater role than any other event in the eventual emergence of the theory of evolution 60 years later.

The discovery of the Rosetta Stone was an epochal event often taken as a metaphor for any major advance in human understanding. A vast field of knowledge—in this case, the entire literature of a civilization spanning 3,000 years—has stood as vast, unknowable, impenetrable mystery. Then, a single key discovery is made that guides the way to total comprehension. The key discovery opens the door, and the remaining work of discovery and interpretation is little more than routine mechanics by comparison. The discovery and appreciation of the true nature of the *Antennapedia* mutation in fruit flies in 1981(Struhl 1981) was a similar event. Once it had been appreciated and decoded to reveal the homeobox and the first *Hox* gene, a relentless process was set in motion that has progressively led to the decipherment of much of the animal world's genetic embryological assembly instructions. This decipherment has greatly clarified not only embryology and the pattern of History of Life, but is now also clarifying the very modular mechanisms by which much of evolution takes place.

When the Rosetta Stone was discovered literally in the backyard of one of one of the French expedition's camps, its importance was appreciated instantly. The Egyptian hieroglyphic language had been a mystery across the ages, and much contemporary wisdom suggested that it was little more than meaningless mystical decoration relating to a ridiculous ancient pagan religion focused on animals and body parts. John Woodward, a professor at Gresham College in London, wrote of the ancient Egyptians in 1777 that "their temples [were] ill conceived and barbarous," and their hieroglyphics were simply "historical representation" of their absurd religious practices. In his 2003 article on the subject, David Boyd Haycock quotes John Montagu, the Earl of Sandwich, who said in 1799 (just before the discovery of the Rosetta Stone) that hieroglyphics were "an invention of the priests who composed them…to make the common people believe that some mystery was couched under them," and that the ancient Egyptian people "also worshipped many ridiculous objects of adoration."

The French scientists believed otherwise. All across the ancient world, in the empire built by Alexander the Great in the fourth century B.C., the Greeks were known to have left "bilinguals"—inscriptions in stone written both in Greek and in the local or "demotic" language. However, no one had ever found a bilingual in Egypt. Pierre Bouchard, an officer in the engineering corps, was sent to demolish some walls for an expansion of a French fort in Rosetta (Rashid). We have no detailed account of the day's events, but work halted abruptly when Bouchard realized he was looking at something no one had ever seen before: a trilingual inscription bearing Greek, demotic, and Egyptian hieroglyphics.

General Jacques Francois Menou immediately declared complete secrecy regarding the discovery, realizing that this was one of the truly momentous scientific events of the millennium. He hurriedly had the Greek portion translated, revealing that it was a message of gratitude from Alexandrian priests addressed to the Greek-appointed ruler, Ptolemy Epiphanes, reciting his work in flood management and his military effort to put down a local rebellion. In an instant, it became almost undeniably clear that hieroglyphics were indeed a form of writing related to an actual language, and that they most likely could and would be deciphered based upon the evidence provided by this most spectacular of archaeological finds. The stone was transported in secrecy to Cairo the following month, and then reported to the world on September 15, 1799, in the official newspaper of the expedition, *La Courier de L'Egypte*. Copies of the stone's inscriptions were made by lithographic technique and sent to academics around the globe to launch the effort at decipherment.

In the mind of the general public, it was Jean-Francois Champollion who accomplished the great task in 1822. (England relied on the famous physicist Thomas Young—as if the hieroglyphics recorded an entirely alien language that could be deduced only by mathematics!) Champollion, however, had learned 12 Middle Eastern languages by age 16, including Coptic, an Egyptian language. Once he realized that the language of the Pharaohs was just another Semitic language similar to Hebrew or Coptic, he rapidly cracked the code, discovering that the writing was a mixture of phonetic and ideographic symbols. It is therefore surprising to learn that academic debate continues about the meaning of a number of hieroglyphic symbols—in particular, the hieroglyphic signs for vertebrae and the spine. This active current debate pertains to the remarkable evidence that the spine played a critical role in classical Egyptian religious thought (Gordon and Schwabe 2004).

An Entrance to the Chamber of the Gods

Geoffroy's ideas advanced tremendously during his journey to Egypt. Most likely this was due to the zoological specimens he found. In addition, his great geographical distance from the formal surroundings and intense competitive pressures of the museum probably helped to further liberate his innate creativity. However, his immersion in the study of ancient Egypt may also have played a role in one of his most fundamental new ideas: While in Egypt, he became fascinated with vertebrae as a common element for understanding the similarity among animals. The interesting possibility is that his vertebral idea may have received some reinforcement and inspiration from the antiquities he was observing. It is not clear how well Geoffroy understood this, but a significant component of ancient Egyptian religion included a focus on the spine and vertebrae. Indeed, two of the most common symbols of the Egyptian religion—the ankh and the djed—are apparently based on the spine.

In July of 1799, a French military victory at Aboukir made it possible for the scientists to journey up the Nile to explore Thebes and a number of other sites of Egyptian temples and tombs. Geoffroy personally removed, opened, and described several human mummies, as well as the mummified remains of many animals. Sadly, important letters from Geoffroy written during his exploration of the Upper Nile are lost forever. He wrote a treatise on Egyptian Theogeny (origin of the gods), but it is unlikely that we will ever know how much his discoveries in antiquities influenced his views of nature. Some of the lost letters were sent on the French ship *l'Osiris*, which sank before reaching France.

Outside of academic Egyptology, discussions of Egyptian gods, temples, and hieroglyphics often seem bizarre and out of place; in this case, however, our key biological theorist was not only present in Egypt when he developed his ideas, but he was also breaking into tombs, carting off mummies, and poring over the Rosetta Stone during a critical three-year period in his intellectual development.

The team of scientists accumulated 45 crates of scientific materials, mummies, statues, research papers, and art before they left Egypt. The contents of those crates are available to the modern reader in the beautiful, 1,000-page, fully illustrated *Description de l'Égypte*. In the recent Taschen republication of this book (France 2002), we see the Rosetta Stone on page 521, Geoffroy's ground-breaking drawing of the skeleton of the lungfish on page 796, and, on no fewer than 17 other pages in the book, we see the Egyptian djed symbol—a representation of the spine of Osiris and a key symbol of the Egyptian religious creation myth. Not only was Geoffroy exposed to this material during his time doing biological research, but his publications are intermingled with publication of the relevant spinal worship symbols.

As many biologists will appreciate, religious concerns often intermingle with ideas about evolution and zoology. It has often been pointed out that if Christian creationism ("Intelligent Design") is to be given equal time with Darwinian Theory (as President George W. Bush and a few school boards have advocated), then we should also give equal time to various other religious views on creation. In general, biologists will feel that no religious creation myths should be discussed in association with a serious scientific work about evolution. However, in this case, it does seem worthwhile to take out a few paragraphs to learn why Geoffroy was seeing Egyptian spinal symbols. In fact, Christian belief and narrative seems to owe a great deal to the ancient Egyptian creation myth.

Every year, for thousands of years in the Egyptian Delta, up until about 300 A.D., the Pharaoh and his priests would raise a huge djed cross. This was done to commemorate the resurrection of Osiris and his role in the immaculate conception of Horus by Isis, wife and sister to Osiris. The djed is a tall vertical with four cross bars near the top, as shown in Figure 3-2 .

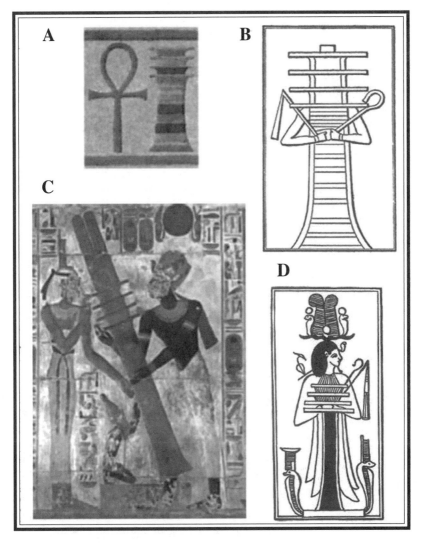

Figure 3-2: The djed symbol.
A) Ankh symbol and djed column.
B) Djed column with arms of Osiris holding crook and flail.
C) Pharaoh Seti I raises the djed column with the help of Isis.
D) Djed column with head of Osiris. Note the head-dress that identifies the role of Osiris in rebirth.

The Egyptian creation myth narrative plays out over a period of millions of years. The Egyptian hieroglyphics that reveal this were an affront to Christian theology during the 19th century, as were the biologists and geologists who claimed that the earth was far older than the 5,000 years officially accepted by the Christian church. (Interestingly, there are hieroglyphic symbols for 100,000 and even 1 million years.) In the narrative, the names and roles of gods as well as the meanings of symbols change over time, and as the story approaches historical times, the gods become more and more similar to living pharaohs. However, much of the detail of the story of Isis and Osiris seems to have been settled by the time of the early Pyramid Texts from the fifth of Egypt's 30 dynasties.

In this myth narrative, there are four siblings: Seth and Osiris, who are brothers, and Isis and Nephthys, who are sisters. Seth becomes jealous when Osiris and Isis also become husband and wife, and he kills Osiris by tricking him into lying down in a beautiful coffin he has made. The body, sealed inside a sarcophagus, is sent out to sea, whereby it arrives in the Lebanese (Phoenician) city of Byblos and becomes incorporated into a tree. A grieving Isis searches the world and finds her lost brother and husband. Disguising herself as a nursemaid, she gains access to the palace where the tree is growing. She reveals herself and is allowed to bring the body home to Egypt. Seth discovers the body and chops it into 14 pieces, which he scatters around Egypt. Isis finds all the parts but one—the generative organ. With the help of magic from the god Thoth, she assembles the disparate parts to complete the resurrection.

Isis bears a son, Horus, by immaculate conception. She is often depicted in religious statuary and paintings sitting with the infant in her lap. Later, she hides him in a basket in the reeds of the Nile so he will not be killed by Seth who now rules Egypt. Horus later grows to avenge his father by battling Seth, and becomes the victorious ruler of Egypt. Osiris becomes lord of the underworld but never returns to normal life. He presides over the court of judgment, where the heart of the deceased is weighed for its righteousness against the feather of truth, leading either to devourment and hell or to immortal union of the body and soul in the Osirian kingdom of the afterlife. Historically, the victory of Horus over Seth is thought to be tied to the warfare involved in the unification of Upper and Lower Egypt (Upper Nile and Delta) that marked a major step the beginning of the Egyptian civilization. Isis worship and the legend of Osiris were important religious elements in the Middle East from 2,500 B.C. until Islam finally obliterated it in 600 A.D.

The Egyptians were greatly concerned with life and death to a degree that probably exceeds that of other cultures, but what was truly unique was their related fascination with body parts. This fascination has been difficult for non-Egyptian societies to fully accept, even at the highest academic levels. According to Sir E.A. Wallis Budge (who published a translation of *The Book of the Dead* in 1895), Isis accomplishes the resurrection of Osiris by reassembling his vertebrae. This parallels a tenet from Egyptian medicine in which it was understood that for a person to stand upright, the spine must be intact. This is clear from the Edwin Smith papyrus, and both the medical and mythological aspects of this were reviewed in some detail recently in the *Journal of Neurosurgery-Spine* (Lang and Kolenda 2002, Filler 2007c). An intact spine is also linked to a key difference between life and death—the vertical and the supine.

When Plutarch wrote his account and historical analysis of the legend of Isis and Osiris, he made no mention of the spine. When Pinturicchio painted a fresco depicting the story of Isis and Osiris on the vault of the audience room of the apartment of Pope Alexander VI in 1492, there was certainly no detail about vertebrae. In a widely published 1998 translation of *The Book of the Dead* by Raymond Faulkner, the mention of the spine is omitted. Nonetheless, the evidence from the artifacts is clear, and it is almost certain that Geoffroy was aware of this. The djed symbol, resembling the spine and ribs, was typically painted on the back of mummy cases and sarcophagi in a way that is unmistakable (see Fig. 3-3, A and B). Two such depictions were published in the *Description de l'Egypte* (see Fig. 3-3, C and D).

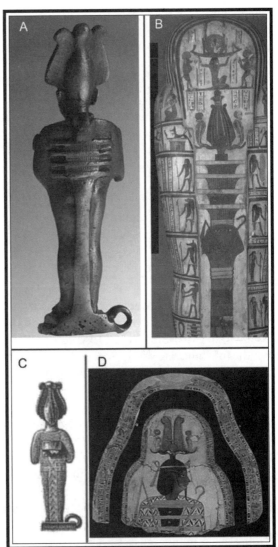

Figure 3-3: Djed icons in spinal configuration.
A) Djed column in position of spinal column and ribs on back of Osiris figurine (by permission of the Virtual Egyptian Museum). B) Djed column with head-dress of Osiris on back of sarcophagus. C) Djed column in position of spinal column and ribs on Osiris figurine. D) Djed column with head of Osiris on back of sarcophagus.

Most informative is Figure 3-3 D. Although the ability to read hieroglyphics was lost for 1,500 years, a good deal of information about Egyptian theology and iconography was maintained through classical times, and was almost certainly well known to Geoffroy and his scientist colleagues. The role of Osiris as lord of the afterlife was known. In addition, Egyptian iconography, in which each of the major gods can be identified in any tableau by their unique identifying head-dress, was well understood. In addition to the head-dress, Osiris (and occasionally the djed column alone, see Figure 3-2 B) is also typically shown with his crook and flail insignia held in his hands. The spinal significance of the djed would likely have been apparent to these scientists, even though the hieroglyphics that later allowed Budge to explain the full details of the link between the vertebral column and the resurrection of Osiris were yet to be deciphered.

The unmistakable publication of the spinal depiction of the djed symbol in the same volume that contained Geoffroy's greatest zoological discovery, *Polypterus*, is more than a coincidence. The contents of this volume were pored over and discussed by this group of scholars in joint seminars for months at a time while the surrounding warfare confined them to their base. Geoffroy himself notoriously collected many of the artifacts, including human mummies and sarcophagi. In fact, Geoffroy stood alone in the final days of the expedition to risk death in order to protect this collection, so it is highly improbable that he was unaware of a symbol that was published on 17 different pages of the summary volume. A man such as Geoffroy, who personally removed mummies and sarcophagi from Egyptian tombs, would have seen the symbols on the back. This is entirely different from the experience of an observer who only sees the coffins faced forward in a museum.

Although Wallis Budge was incredibly prolific (he wrote more than 50 books, most still available today), his work has been subject to constant attack from within the profession of Egyptology. To start with, his origins are problematic for some: He was born to a poor, unmarried young woman in Cornwall, fathered by an unknown waiter in a hotel. Wallis Budge's interests and abilities in languages were so striking that friends managed to raise funds to send him to Cambridge. Nonetheless, his class, accent, and entertaining personal manner no doubt earned him the disdain of the highborn classical scholars whom he so magnificently outshone. Some modern scholars have followed a tradition of either attacking this early giant of their field or simply questioning him because of the supposed progress in the 100 years since he published his translation of *The Book of the Dead*. Nonetheless, he is widely respected in Egyptology, and many of his conclusions enjoy ongoing support by many in his field. Budge's translations are still the standard. In particular, the hieroglyphic that is numbered as F37 in A.H. Gardiner's system does in fact appear to refer to vertebrae. A recent book covers the subject in considerable detail (Gordon and Schwabe 2004). This is also the hieroglyphic used in the description of the resurrection in Chapter 50 of *The Book of the Dead*.

What has Raymond Faulkner done in removing the anatomic references in his new translation? The idea is clear when you look at another phrase in this section of the text. The Budge translation reads "rising up from my weakness upon my two legs." The hieroglyphic symbol used here actually includes a drawing of two legs. However, Faulkner translates the phrase as "when he caused me to become hale." He is telling us what he believes the passage means rather than really translating the Egyptian. An approving preface to the translation is written by Carol Andrews, yet Andrews recites the spinal interpretation in her own book (Andrews 2004). In fact, a mummy case with the djed symbol along its back in the position of spine and ribs is used to illustrate the back cover of her book (see Figure 3-3 B). In her book *Egyptian Mythology*, Geraldine Pinch, a faculty member at Oxford and editor of several academic Egyptology journals, also accepts the vertebral explanation of the Osirian djed symbol (Pinch 2004).

When an Egyptian was buried, symbols of the spine of Osiris were considered crucial to the opportunity for resurrection in the afterlife. The resurrection of Osiris through the assembly of his vertebrae by Isis was described in the funerary texts placed alongside most major Egyptian burials for thousands of years. The raising of the djed column commemorates this, and, as many religious authorities have pointed out, Christ's resurrection and the Roman Cross make much more sense when it is appreciated that for thousands of years in the Middle East, erection of the djed cross was the symbol of resurrection of Osiris, king of the afterlife.

Neither the Hebrew religion nor the Roman religions of the early centuries had a significant formal orientation toward an afterlife. In Judaism, the individual lived "dust to dust," and in the various Roman traditions there was no clear conviction that any form of immortality existed for the ordinary person after death. There was certainly no well-developed focus on a heaven and hell, or upon an assignment to one or the other destination based on one's behavior or belief during life. These beliefs were features of Egyptian religion as it was practiced for thousands of years, up to about 300 A.D., and of Christianity as it emerged toward the end of Egyptian civilization in the first several centuries A.D.

The connections between various religions are difficult to trace. The ancient Hindu Vedas of 1,500 B.C. have a concept of a series of chakras stacked up vertically that constitute the various elements of the soul, but it there is no definitely proven link to the Egyptian spinal focus as embodied in the djed column. Nonetheless, it is clear that the early Coptic Christians of Egypt were significantly influenced by ancient Egyptian religious traditions, and that Coptic beliefs in turn had an important impact on Constantine when the formal aspects of Christianity were codified in the fourth century A.D. It seems quite likely that early Christians appropriated not only the symbol of the Osirian djed cross in relation to resurrection, but also the concept of the afterlife and of judgment at the time of death (after the Egyptian concept of the weighing of the heart to determine who would be granted eternal after-life). Of course, the idea and icon of Isis and her infant born by immaculate conception was another dominant symbol available to the early Christian theologians.

Although this may be far more than you wanted to know about Egyptian religion, it is certainly refreshing to have a zoologist linked to an exploration of the origins of Christianity, rather than have a theologically driven Christian exploring the origins of evolutionary theory! This all raises the interesting proposition that in that dockside encampment in Alexandria on October 1, 1801, Étienne Geoffroy Saint-Hilaire had in hand clues to a unified theory of explanation for similarity among animals, as well as the information (the Rosetta Stone, *The Book of the Dead*, pharaonic religious symbols, and so on) to explain the Egyptian origin of some of the major symbols that later evolved into the symbols of the Hebrew and Christian faiths (the cross, the resurrection, the afterlife, the immaculate conception, mother and child, Moses in the reeds, and so on). Little wonder, then, that Geoffroy was prepared to die that day rather than give up his scientific treasures!

A Final Conflict and an End to the Journey

As the five-month British siege of Alexandria in 1801 finally broke the resistance of the last remaining French outpost of the Napoleonic invasion, the British demanded that French scientists turn over to them everything they had collected, drawn, or written. They would be allowed safe passage back to France along with the scientific instruments and the books they had brought with them originally, but all of the spectacular results of their three years of systematic research into the history, architecture, zoology, botany, and culture of Egypt must be surrendered and transported to Britain.

The French general Menou apparently was most concerned about keeping the Rosetta Stone, as well as another treasure that had been kept hidden and secret; he cared little for the scientist's work. One French scientist, Savigny, suggested that French scientists accompany their scientific treasures to London, where they could complete their work, but this was not what the British had in mind, and so the offer was turned down. In any case, General Menou had stated that any French scientists who wanted to do this would need to be stuffed, wrapped, and crated up themselves if they wanted to go to England. As the British slowly advanced from Aboukir Bay toward Alexandria, Menou held out for nearly two months solely over the issue of control of the antiquities.

The British commander John Hely Hutchinson did not really know what the French had collected. It probably made little sense to him that General Menou continued to resist on this point alone. He sent the diplomat William Richard Hamilton, along with Edward Daniel Clarke (a Cambridge mineralogist and traveler) and John Marten Cripps (Clarke's student and financial backer), to catalog and take charge of the scientific materials. This trio had originally brought news to Hutchinson of the general extent of the French treasures. Eventually Clarke learned that the French believed they had found the sarcophagus of Alexander the Great and that this seemed to be the prized possession. Clarke located it hidden in the hold of a French hospital ship under bandages and filth and brought it to Hutchinson.

Because Clarke's claim that he had found the Tomb of Alexander has always been doubted, and because Clarke himself was really a young traveler on a two-year expedition to see the world and sell stories of his exploits, this particular detail tends to be omitted from the story of the French expedition. Ever since the supposed recovery of Alexander's tomb by Clarke, several major expeditions have been dispatched to the area to find the actual final tomb. However, recent scholarship shows that what Clarke discovered probably was indeed Alexander's final sarcophagus (Chug 2002). It is still at the British Museum today. If there was any treasure greater than the Rosetta Stone—at least from the point of view of Napoleon Bonaparte—there can be no doubt that it was Alexander's tomb. The French fleet was sunk within two weeks of Napoleon's original arrival in Egypt, so once the sarcophagus was found it could not be carried back to France. This seems to explain why Menou finally gave in. Hundreds of French soldiers were being killed while he held out apparently over the antiquities and the scientific data. Yet, when Clarke and Hamilton visited Menou at his quarters with news of their discovery on the hospital ship, Menou told them they could have the Rosetta Stone and what ever else they wanted.

Geoffroy apparently appreciated that it was Hamilton and Clarke who were driving the demands from General Hutchinson to get control of the scientists' materials. When Geoffroy learned from them that General Menou had finally agreed to turn over the Rosetta Stone, he realized that the military would no longer protect his collections. Isidore Geoffroy, in his 1847 biography of his famous father, reports that an enraged Geoffroy menaced a pale and silent Hamilton, who was stunned to hear the scientist's threat. Geoffroy pointed out that the nearest British soldiers were still two days away and shouted that the French scientists would not obey the agreement between

Hutchinson and Menou: "We will burn our riches ourselves…. Then you can do with us what you want…. If it's fame you want," he thundered, "count on the memory of history: you too will have burnt a library in Alexandria!"

When Omar, Calif of Baghdad, captured Alexandria in the name of Islam in 642 AD., forced conversion of the populace to Islam was not enough. The library at Alexandria was one of the wonders of the ancient world. It was the only truly complete library in its time and dwarfed every other collection with its thousands upon thousands of scrolls from Rome, Egypt, Greece, Persia, Africa, and the rest of the known world—literature, religious works, history, and philosophy. As it was all written by hand, most of these were the sole extant copies. Omar decreed that every single item in the library—excepting only the works of Aristotle—should be systematically burned. Indeed, only some of Aristotle's notes survived the burning of the library at Alexandria. This was and still is widely regarded as one the greatest intellectual catastrophes in human history.

Clarke, who later became a professor at Cambridge, apparently judged his odds and reported back to Hutchinson that he must let the French scientists leave with their works. Hutchinson annulled the part of the agreement requiring the surrender of the documents, collected the Rosetta Stone, and allowed Geoffroy and his colleagues to depart for France with their vast collections. Clarke later published a report on the purported sarcophagus of Alexander. He avoided being tarred with memory of Omar's great intellectual crime, but he never really convinced the contemporary world that he had seized the correct sarcophagus. Did Geoffroy really intimidate the British with his threat? Or rather, did Hutchinson—having seized control of both the Rosetta Stone and Alexander's sarcophagus—now feel that he'd done enough?

As Isidore Geoffroy puts it: "This was the last event of the Expedition of Egypt; and as the history of France is written, so brilliant a start, so sad a conclusion; however, thanks to our scientists it will all be remembered as a National glory!" And indeed, Geoffroy's determined stance on the docks made him an instant national hero and gained him the award from Napoleon—the cross of the French Legion of Honor. Although this award is given frequently in France today, Napoleon had just created the Legion, and an award to a scientist for risking his life by defying the victorious British Army in defense of his data is a bit of a landmark in modern science.

The Significance of the Data

Writing to Cuvier in August of 1799, just prior to the expedition to the Upper Nile, Geoffroy had become extremely excited by his biological findings in the unusual fish he had been collecting: "I require nothing less than the Throne of Anatomy. If you hesitate, I retort: have you ever found in one fish the organization of the quadrupeds and of the squid?" He describes in considerable detail the internal anatomy of the air-breathing catfish with a three-chambered heart (*Heterobranchus*), which he initially called *Silurus anguillarus*. Later he described *Polypterus* as a previously unknown and primitive type of fish with cartilage in place of bone, possessing not only lungs and the ability to walk on land, but also a jaw similar to those of tetrapod (quadruped) land animals, rather than the complex, specialized collection of jaw elements found in the vast majority of fish.

He conceived of one of the cornerstones of modern systematic biology when he had the idea of establishing the homology of bones by looking to their surrounding connections. Whether or not a bone closely resembled the same bone in another species, you could demonstrate their similarity if they each had the came pattern of connections to surrounding structures. He also established a second cornerstone when he pointed out that the *centers of ossification* in embryos (the points where the cartilage begins to convert to bone) could reveal homologies that were obscured when two bones fused into one in the adult animal. This is different from the more controversial idea of others that "ontogeny recapitulates phylogeny." That concept implies that the embryo actually takes the form of more ancestral embryos as it progresses to completion as a more *derived*, or further evolved, descendant. Geoffroy was pointing out that studying embryos could reveal underlying similarities between two species that were more difficult to detect or confirm in the modified and fully developed anatomy of the adult.

These two principles—the use connectivity and the use of embryonic ossification centers—helped clarify many similarities and differences among animals and were the basis of many important advances in zoology in the following two hundred years. These have been accepted as Geoffroy's most enduring contributions to biological science. Now we realize his even grander contribution in giving us the theory of unity of animal life as defined by serial body segments. These findings led Geoffroy to propose that a single plan for bony anatomy could be applied not only to birds, mammals, reptiles, and amphibians, but to fish as well. By the time Geoffroy left Egypt, he seems to have come to the conclusion that, in the repeating structure of the vertebrae of animals, there was some clue to the unity of design among all creatures. In the 1990s, it was discovered that the *Hox* genes that determine the segmentation of fruit flies are virtually identical to the genes that determine the organization of human segments. This is what has raised Geoffroy's ideas to modern prominence. Similar to Mendel, who discovered the foundations of genetics before anything was known about the existence of genes, Geoffroy detailed the role of segmented structures in the unity of animal life nearly two centuries before the homeotic gene system was discovered.

Reestablishing His Place After a Return to France

When Geoffroy returned to France in 1801, he found that although Cuvier had been promoted to a coveted position in the French Academy of Science, the risks, the bravery, and the discoveries of Geoffroy's own Egyptian years meant nothing for scientific advancement. Perhaps out of jealousy from those who had never done more than travel to their office to describe specimens, Geoffroy found that all his colleagues could see in his expedition was nearly four years away from his post. He got appreciation only for the specimens he had brought back for them to work on.

Further, in the final weeks of the British siege against French forces in Alexandria, Geoffroy was working feverishly on a grand unified theory of animal electricity, precipitated by his discovery of new species of electric eels. Upon his return to France, he was gravely warned to be as formal and traditional as possible in his scientific publications. The theory of electricity must never see the light of day. Cuvier, in Geoffroy's absence, had achieved high scientific office by success as the most conservative and careful of scientists. If Geoffroy wanted success in French science, he would need to restrict his work to the kind of strictly objective detailed work that Cuvier had now established as a standard. Cuvier, as Geoffroy rapidly learned, was no longer his friend.

Nonetheless, by 1807 Geoffroy produced a series of publications that showed what he had seen and understood in Egypt. Further, he outlined his concept of unity of plan and unity of composition, showed his new methodology, and proved his point regarding fish and other vertebrates. Because he now had a program of his own and had made major advances that were published in the proper, formal way, he succeeded that year in advancement to the Academy of Science alongside Cuvier.

The Conflict of Ideas

Aristotle had divided all of the animals into two large categories: animals with blood (mammals, birds, reptiles, amphibians, fish) and animals without blood (the remaining insects, crustaceans, and worms). In the 18th century, zoologists updated Aristotle by changing the terminology to "animals with red blood" and "animals with white blood." In 1794, in his first lecture at the newly formed National Museum of Natural History in Paris, Geoffroy's colleague Lamarck had proposed changing the terminology to vertebrates and invertebrates—probably Lamarck's most successful and enduring contribution. Cuvier was further dividing the animal kingdom into four entirely separate categories (or *embranchements*).

In 1822, Geoffroy extended his claim for unity of plan to include the insects and crustaceans along with the vertebrates. In doing so, he was taking animal categorization in the opposite direction from Cuvier and confounding the definition proposed by Lamarck. Finally, after introducing a presentation about mollusks by some young scholars at the museum in 1830, Geoffroy extended his claim of unity of plan to include the cephalopods (octopus and squid). The idea was that an octopus was effectively similar to any other model animal in Geoffroy's plan, but bent over in half so that the arms and legs were next to each other. Recent evidence from the study of *Hox* genes suggests that Geoffroy appears to have been correct on this point as well (Lee et al. 2003). However, after quietly responding to Geoffroy for 35 years, Cuvier finally decided at this point that it was time to attack Geoffroy's ideas publicly and directly. A public, formal, and published exchange of positions and views, carried out at regular meetings over a two-month period, constituted the Academy Debate of 1830 that drew so much attention.

Cuvier's view can be understood as incorporating two important threads of biological science. First and foremost, he felt that scientific hypotheses should be closely limited to the specific data at hand. With regard to the various types of animals, he felt that they represented very different independent acts of creation. Along with Aristotle and Darwin, he also believed that the most productive way to understand the anatomy of an animal was to look at the function of each component of the animal. From Cuvier's point of view, this presumes that God made every individual part of every animal for a good reason; the anatomy appears the way it does because, in God's view, that was the best way for that animal to accomplish the functions it needed in order to succeed in its environment.

Geoffroy's view differs from Cuvier in that he believes that science moves forward by sweeping intuitive interpretations, and that the collection of specific data should support and fill in details. However, he does not expect to limit the generation of scientific theories to the results of individual findings. For instance, when you find for the first time a fish that has lungs, can breathe in the air, has a jaw capable of biting rather than just sucking water, and is able to pull itself along on land with its fins, you should not be limited to commenting about the special capabilities of one odd species of fish. Rather, it is fair to begin with the idea that there will be a connection between quadrupedal land animals and the fish of the sea, and then use the lungfish to support the idea of a connection. Cuvier is happy to classify *Polypterus* relative to other fish and move on to the next species of trout that is discovered. In contrast, Geoffroy sees in this fish a clue to a grand pattern of nature.

Geoffroy's other principal point of difference is that he looks to a pattern of underlying biological rules to explain the anatomy of a creature, rather than looking for an optimized function for each individual part of the animal to explain the design of the whole. For example, we can consider the limb of a modern horse and that of an extinct, ancestral horse species of the past. We know that an animal can run perfectly well whether it has five toes on its feet (primate, rat, lion) or whether it runs on the toenails of one or two toes (horse, antelope). We look at the leg of the horse and appreciate that, in addition to the main central toe bones, there are small toe bones on either side that don't reach the ground. Geoffroy would say that the smaller side bones reflect the inherent ancestral anatomy of five digits in the foot, which is why the two smaller, useless toes are seen. Cuvier would say that there is a reason the horse has the two smaller side toes, and that we need only discover the function as we learn more about the animal. After all, why would God put the extra toes there if they didn't have a function? Further, Geoffroy asserted that a given structure in two different species could be homologous with each other even though they had completely different functions. An example would be a bone that served as part of a vertebra or the shoulder bones in one species, but is fused into a skull bone in another species. This is a fundamental concept that has been another cornerstone of modern biology—namely, that we can trace the fundamental identity of a given anatomical or biochemical structure across wide divides in the animal kingdom, even when the structure has changed its shape and function.

In a statement that is incredibly prophetic in relation to what we now know about the protean applications of *Hox* genes and the wide variety of other anatomical structures studied by zoologists, Geoffroy wrote in one of his 1807 papers:

> Nature constantly uses the same materials, and is ingenious only in varying their forms. As if in fact she were constrained by initial givens, we see her always tending to make the same elements reappear, in the same number, in the same circumstances, and with the same connections.... But all are nonetheless conserved, although in a minimal degree that often leaves them without utility: they become, as it were, so many rudiments that testify in some sort to the permanence of the general plan.

It was objected that this position was offensive from a religious point of view. It seems to be showing that God is restricted in how he can construct an animal. He is being forced to use the same parts over and over again. Does Geoffroy mean to say that someone or something is forbidding God from making up a new bone when he wants to? Geoffroy just felt that this is what the data showed. Later in the century, Sir Richard Owen argued that these restrictions should still be considered as admirable signs of an unknown divine plan. And it was Owen who actually established the meaning of the term *homology*—the fundamental similarity in parts of two organisms—as we know it today. Geoffroy however seemed unconcerned by the theological implications of where his science was leading him.

Before the publication of Darwin's *The Origin of Species*, the fact that every feature of an animal seemed to have a specific purpose for which it had been designed seemed to constitute evidence of God's hand. This view—also called teleology—is behind the religious pseudo-science of "intelligent design." Darwin substituted natural selection for the deity-as-designer, and accepted some underlying similarities of animals due to common ancestry across time, rather than due to the "habit" of an intelligent designer to make animals in similar ways. Geoffroy's ideas place the focus instead upon understanding the constraints that maintain the unity of design despite the variably significant impacts of functional adaptation.

Geoffroy's scheme suggests that whether we look to the adult or the embryo, and whether we are looking at a mammal or a fish, we should generally be able to find every one of the bones and organs that constitute the fundamental animal plan. This is the unity of plan he is advocating. When he first wrote about the subject in 1795, he felt there would be a separate plan for each class of animals—one for mammals, one for birds, one for fish, and so on. When he returned from Egypt, he was armed with his new methods of homology from connections and the use of embryology to reveal hidden homology. He was convinced that using these new tools, together with analysis of the spectacular new types of fish he had brought home to France, he would be able to find the bones of the quadrupedal land animals in the skeletons of the fish and unite them all in one plan.

In his descriptions of fish vertebrae in his letters from Egypt to Cuvier (see Geoffroy and Hamy 1901), we can see that he was already considering the idea of linking the crustaceans (lobsters), arachnids (spiders), and hexapods (insects—modern: phylum *Arthropoda*) together with the vertebrate animals. The possibility grew out of his observation of blood vessels running along the body inside an enclosed channel extending from the vertebrae, just as the spinal cord was in its own bony channel along the vertebral column. He explains this in more detail in his 1822 report, "General Considerations on the Vertebra," in which he also explains that he first presented the ideas in public in an open lecture in 1809, and had previously published them in the journal *Isis*.

The Inverted Insect and the Origin of Vertebrates

When he formally presented his concept of inclusion of the arthropods with the vertebrates in 1822, Geoffroy argued that the external body segments of the arthropods were effectively the outer walls of their vertebrae and that all of the organs were inside the vertebrae, unlike the limited numbers of organs inside the vertebrae of vertebrates. He applied his principal of connection by pointing out the well-known fact that, in arthropods, the muscles attach outward to the body wall, rather than inward to the bones as in vertebrates.

An Upside-Down Arthropod

Another major problem in linking the arthropods to the vertebrates is that, in arthropods, the nerve cord runs along the front of the animal facing the ground, with the digestive track along the back facing the sun. Vertebrates have the opposite arrangement, with the nerve cord at the back and the digestive tract toward the front. In his 1822 paper on the subject, Geoffroy points out that from the point of view of the body plan, it doesn't matter if a creature walks with its stomach facing the sun or facing the ground—everything inside can still fit the plan if you just flip the body over. Genetic confirmation of the identity between insect segments and vertebrate segments had already rocketed Geoffroy back into the current scientific literature in the 1980s, but the subsequent genetic confirmation of his "dorso-ventral" flip in the 1990s helped put him into college biology textbooks as well.

An explanation for crossed nerve tracts

Because the components of the head seem similarly oriented in vertebrates and arthropods, Geoffroy proposed another interesting anatomical confirmation of the event. As was well known at the time, in vertebrate animals, the right side of the brain controls the left side of the body. This is accomplished through a *decussation*, or crossing, of a majority of nerves descending from the brain at the level of the hindbrain or medulla. Geoffroy has provided really the only scientific suggestion to explain this odd fact. In the 1822 paper, he argues that the body of vertebrates has been flipped over while the head has remained in the same position that it has in arthropods, and that this is the cause for the decussation of the nerve tracts.

This idea has proven reasonably convincing because we now know that the genetic determinants of basic brain formation are identical in vertebrates and in arthropods. Some debate continues, but it seems increasingly clear that the brain arose only one time in the History of Life on earth, and that this occurred prior to the separation of these two lineages. A paper on this topic by Reichert and Simeone (Reichert and Simeone 2001) quotes a large number of papers from 1991 onward, as well as one lone paper from 1822—Geoffroy's now-famous article. As has been pointed out by Hervé Le Guyader in his review and republication of many of Geoffroy's key works (*A Visionary Naturalist*, published in 1998 in French and 2004 in English), there seem to be virtually no primary academic works published on this subject between 1822 and 1991.

The implications for systematics and zoology

At the time of Geoffroy's 1822 publication, Cuvier became the first to heap ridicule on the idea: "Nothing in common, absolutely nothing, between the insects and the vertebrates; at the very most one single point, animality." Now we understand that a surprisingly large number of the genes that control early development and assembly in organisms are in fact nearly identical between insects and vertebrates, and that many structures that seem to be formally similar or analogous are actually homologous from the point of view of the genetic control of their assembly.

Aside from the stunning breach this creates in our long held Aristotelian view of the organization of the animal world, the mechanics of the inversion event pose a challenging problem for Evolutionary Theory. To the extent that the genetic control of embryo formation is understood today, we know that a great deal depends on the creation of gradients (the steady variation of concentration) of chemical signals across the tissues of the forming embryo. Various genes check the local environment around them, determine the strength of the surrounding gradient, and are thus directed how to act and what tissues to form. It is the readout of the dorso-ventral body axis that is flipped in the origin of the vertebrate lineage. Formation of the head is under control of a different set of genes from those of the body, and so may not have been subject to the same flip.

The genetic and evolutionary mechanics of the dorso-ventral flip

This change in dorso-ventral organization appears to be coded for by a very simple genetic change, causing the gradient to be read out in the opposite direction from its normal sense. However, this does mean that we must accommodate a transformation from an invertebrate body organization to a vertebrate body organization very rapidly. In fact, the whole event has to happen in a single sudden mutation in a single generation.

Evolution as we understand it depends significantly upon speciation events. When a new species arises, it generally cannot interbreed successfully with members of its parent species. If the two can interbreed, the two groups eventually fuse back together into a single population. We also understand that there are a variety of methods by which speciation can occur. As described in Chapter 5, there is one type of speciation event—chromosomal speciation—that can take place over just two or three generations. It does not depend on the presence or absence of any differences in the external anatomy of the two groups, and it does not require that they be separated by any physical barrier. Because of the genetic mechanics of the dorso-ventral flip, it is entirely compatible with a chromosomal speciation event.

Therefore we imagine these odd, inverted, wormlike creatures living in the sea among other individuals of their species who are not inverted. Cross fertility is maintained initially in the middle steps of chromosomal speciation as the number of these flipped descendants increases, and then a second chromosomal event seals off the new species. In this fashion we have sudden generation of a novel body plan. The only input from natural selection is that these odd mutants have to be able to survive long enough to reproduce while living in their environment and amongst the individuals of the parent species.

Anterior-Posterior, Oral-Aboral— Yet Another Flip

Terminology and classification schemes also appear to have played an important role in explaining the reluctance of biologists to consider Geoffroy's idea. Aristotle's scheme of dividing the animal kingdom into blooded and non-blooded animals, and Lamarck's division of vertebrates and invertebrates produce some conflicts with each other, but despite this they do tend to direct attention to the key features of the circulatory system and the vertebrate spine.

Most modern biology texts have divided the animal kingdom by use of a different distinction, one suggested in 1908 by Karl Grobben, an invertebrate zoologist at the University of Vienna. It is a technical but very fundamental difference that distinguishes the early embryological development of animals into two large groups. This is the distinction between *deuterostomes* and *protostomes*, and it relates to a critical event early in embryogenesis.

At an early stage of development of most animals there is a ball of cells. At one point on the ball, the surface begins to dip inward toward the center. As this continues, the inward dip becomes a tube leading from the outside of the ball into the interior. This is the origin of the digestive tract. In protostomes, the point of the inward dip eventually becomes the mouth. However, in deuterostomes, the point of the initial inward dip becomes the anus, and a second opening must form later to become the mouth. This deuterostome (literally "second mouth") group includes the vertebrates (see Figure 7-1).

This major distinction among the animals holds up fairly well when tested by genetic analyses of relatedness. The best example here is the starfish: A starfish doesn't have anything that remotely resembles a vertebral column or even a proper notochord (the primitive linear structure that runs from head to tail in virtually all members of the chordate animal group, including the vertebrates). It does not have blood, red or otherwise. However, embryologically it is a deuterostome, and, sure enough, its DNA shows that it is closer to the vertebrate lineage than any of the protostomes.

Geoffroy's idea seems to disregard all of these concepts when he looks for a common body plan between vertebrates and arthropods. The arthropods don't have blood per se—at least not oxygen-carrying red blood—and they don't have vertebral columns. The dorso-ventral flip he proposes doesn't seem to address the problem of protostome vs. deuterostome because it doesn't seem to explain how the digestive tract gets turned around from front to back. In fact, Geoffroy's theory *does* accommodate this issue. For his dorso-ventral flip of the body, we can see two outward manifestations: 1) the nerve cord faces the sun, whereas the digestive system faces the ground; and 2) the dorso-ventral orientation of the head relative to the rest of the body has been flipped so that we see the decussation of the nerve fibers. However, if the entire anterior-posterior orientation is flipped with regard to the embryonic ball, there is not necessarily any external manifestation of this in the adult animal. Only in the early embryo is the anterior-posterior flip apparent.

There is a genetic embryological clue that might help, but we still don't have enough adequate information to test it. Deuterostome animals have a developmental gene called *Nodal* that determines the left side of the embryo as opposed to the right. This helps establish the right-left gradient. If *Nodal* existed in Protostomes, we might expect to find it on the right side instead if there were just a dorso-ventral flip, but on the left if there occurred a second anterior-posterior flip. However, protostomes have not yet been shown to have this *Nodal* protein function.

It is also worthwhile to note that the dorso-ventral flip and the protostome-deuterostome change need not have occurred at the same time. The deuterostome group has three major types of animals in it: the starfish/sea cucumber group, the acorn worms, and the chordates. The starfish (*echinodermata*) and the acorn worms (*hemichordata*) don't really have anything similar to a nerve cord, and their overall body organization is so different from chordates that the organization of their dorso-ventral body axis is difficult to correlate with the structure in chordates. The situation can probably be resolved with more genetic analysis, but it is certainly possible that the anterior-posterior flip that sets up the mouth/anus development issue took place first, and that the dorso-ventral flip of the body happened later, but only in the ancestor of the chordates.

Are Vertebrates Inside Out?

Finally, we come to the issue of how the arthropods have most of their body tissues inside with their "vertebrae" outside, whereas vertebrates have their vertebrae on the inside. Does this mean that vertebrates are more or less turned inside out relative to arthropods? This is one of the issues that was raised to refute Geoffroy's idea. The hard outer body wall of arthropods (such as that found on insects) is not related embryologically to the type of tissue out of which vertebrae are fashioned. Despite the differences in embryological origin between the hard exoskeleton on the outside in one group as opposed to the hard endoskeleton inside the other group, Geoffroy felt there was a fundamental homology between the two tissues because of his law of connections. The body muscles attach to the external body wall in arthropods just as they attach to the vertebrae inside in vertebrates. Interestingly enough, the homeotic genes are expressed in the ectoderm or outer skin of arthropods but, with the exception of the nerve tissues, they are not expressed in the outer skin of vertebrates but rather in the internal structures that include the vertebrae.

The result of all of these considerations is that the emergence of the clade (or lineage) that includes the vertebrates may have undergone a flip of its dorso-ventral axis, of its antero-posterior axis, and even an inside-out flip of tissues. If all of these embryological axes flipped, it is certainly remarkable that the progeny were able to survive, reproduce, and generate our lineage.

Modern Scientific Objections

One of the modern attacks on the concept of unity of plan comes from scientists such as Ronald Jenner who threaten to tar the new advocates with the label of "typology." A similar concern has been expressed in the past by Ernst Mayr. The concept of ideal type or archetype has roots in Greek philosophy and theology. In its purest form it follows the biblical dictum that God made man in his image. Thus, humans were designed to reflect the appearance of God, and the other animals are just more or less degenerate copies of the human form. Scientists before Darwin were often interested in proposing a purely zoological archetype, rather than using man or God as the model. Any animal form could then be generated from this basic common design.

Despite the religious links, many scientists believed that such an archetype could be a purely biological phenomenon independent of any religious phenomena. Darwin's substitution of the shared common ancestor for the shared archetypal model was one such concept that could be similarly tarred with the label of "typology," although it is widely accepted as scientific and not metaphysical. Most of us understand the utility of the periodic table of elements. There are subatomic particles that are glossed over at the level of elements, and there is a nearly infinite variety of different chemicals, substances, stellar bodies, and organisms to be studied that are made up of elements. As with the elements of the periodic table that first inspired Goethe and Geoffroy, there are fundamental components in biology as well. Understanding these components and modules is appropriate and necessary because that seems to be how nature is organized.

As a number of evo-devo specialists have pointed out, Jenner's concerns seem to be focused on terminology. The vast array of detail we now have about the genetic control of embryogenesis is limited to data on a very small number of well-studied species: a few mammals, a few insects, one species of bird, one fish, and a few other invertebrate species. These species are used as examples to signify the general process of embryological development among very large numbers of related species. Inferences are then made about genes in common ancestors, even though we can never actually study the genes of those individual ancestors.

However, stable modules do exist in biology without being metaphysical aberrations from science. As Carl Woese has shown, the key biological module of transcription/translation of DNA to protein is incredibly ancient, and the differences in detail in the mechanism are relatively minor across huge segments of the living world (Woese 2002). Knowing a key derived anatomical feature for every organism we classify is essential. However, we may gain very little knowledge about biological mechanisms by trying to fully analyze the minor variations in the transcription/translation module in every one of millions of different species. According to T. W. Deacon in his 1997 book, *The Symbolic Species*, generalization is not only a fundamental aspect of human intelligence: It is the basis of most of our modern science outside of cladistics (a modern system for studying relationships among species). We are not making a metaphysical error if we try to understand the basic workings of the cosmos without first carrying out a plan to fully describe every planet and star in the universe.

Étienne Geoffroy was not concerned with the concept of archetype. He was doing what a biologist should do: looking for clues to explain the structure of the biotic world. Further, it is ridiculous for any modern scientist to suggest that the pursuit of commonality of animal design has not been spectacularly productive of some of the finest science of the 20th and early 21st century. When the objective observer looks on the past 20 years of literature in morphological genetics, homeobox genes, and *Hox* gene family evolution, there can be no doubt that we have at hand one of the greatest accomplishments science has ever achieved: a decipherment of the genetic instructions for the building of animal bodies.

It is all the more stunning to consider that Geoffroy's 1822 publication represented an act of considerable personal and intellectual bravery. He was not intimidated by the fear of being incorrectly tarred with the metaphysically tainted label of typology. Perhaps because of a lack of similar determination on the part of other scientists, research in this critical aspect of biology came to a halt. This is why a period of 170 years elapsed in which not one single significant publication appeared in this utterly essential field.

There is however one absolutely fundamental aspect of Geoffroy's work in which most of modern science sides with Cuvier. This is the apparent imbalance between theory and evidence in his work. By studying ossification centers in juvenile specimens he collected, and by applying his now widely accepted rules of what we now call homology, Geoffroy did a creditable job of proving the commonality of skeletal design between fish and the quadrupedal land vertebrates. However, it is not clear that he had any evidence at all for his theory that united the arthropods and the vertebrates. Was he correct by chance, or does science occasionally require grand visions from well-informed scientists in order to map out the way forward? Did Cuvier's more conservative approach provide a vast catalogue of descriptive detail with no effective theoretical value? At the outset of this chapter, Geoffroy's career was tied to the birth of modern biology. What Geoffroy did was to lay out the project of explaining similarity of animal design, map out the major problem to be solved, design useful methodology that has remained in use to this day, go out into the field to seek and analyze data from the biological world in support of his project, then boldly publish both the detail and the theory that resulted from his efforts.

Work on these questions today is progressing towards a sort of biological grand unification. From a number of perspectives, this effort mirrors similar efforts in other fields, such as physics and cosmology. One of the great paradoxes of animal biology in the 1970s and 80s was why animal anatomies were so utterly different from phylum to phylum when their biochemistry and cellular processes were so remarkably similar. By showing how to see the underlying similarities, Geoffroy's program has led the way toward understanding common patterns in animal form, as well as at the level of biochemical and cellular processes.

How Species Originate

Introduction

It is possible for one species to turn into a different species. This is called *transmutation*. Some kind of mechanism is required in order to change a species, and an additional mechanism is required to actually separate it as a new and distinct species. The process of generating a new species is called *speciation*. When a sudden major change of anatomy takes place due to a mutation, this will lead to the formation of a new species only if a speciation event also takes place. The occurrence of change due to mutation is easy for most people to understand and accept, as there are countless examples of this both in science and medicine. What has proven far more difficult for some to accept is the potential for speciation to occur as well.

Speciation: How One Thing Can Change Into Another

Transmutation: An Example From Alchemy and Nuclear Physics

In many ways, the study of evolution is all about how one thing changes into something else. We can get a useful handle on the issues by considering a famous story from physics and chemistry that took place at the beginning of the 20th century. The alchemists of the Middle Ages believed they could convert "base metals" (for example, lead or copper) into gold. Goethe's protagonist, Faust, himself is deeply involved in this. However, after Lavoisier revealed the nature of the elements in the 1760s and Mendeleev proposed the periodic table of elements in the 1869, the idea of transmutation of one element into another was relegated to the junk heap of magic and mysticism.

In fact, when 31-year-old physicist Ernest Rutherford reported in 1902 that thorium could undergo "spontaneous transformation" into another element, he was widely denounced for promoting alchemy. Just four years earlier, French physicist Henri Becquerel had discovered the emanation of radiation from uranium. A growing number of scientists were investigating the strange new physics of uranium decay when Rutherford published his interpretation of what he had observed in his Montreal laboratory at McGill University (Rutherford and Soddy 1902). Within a few years there was virtually no disagreement among physicists that transformation in nature was a reality.

The next major shock to the sense of constancy in the world was set in motion in 1918 when the Austrian physicist Lise Meitner, working with German chemist Otto Hahn, identified a new element, protactinium, that had one less neutron than uranium. 20 years later, they tried to create an entirely new element, with one neutron more than uranium. The plan was to try to accomplish this by bombarding uranium with neutrons. Chemical analysis of the results seemed to indicate that they had failed. Inexplicably, all that could be found afterward were some small contaminating amounts of totally unrelated small atoms of krypton and barium. Hahn was disappointed, but Meitner—who was fleeing Nazi Germany in 1939 when she heard the results of her experiments—had a different interpretation. She noticed that the atomic number of krypton (92) added to the atomic number of barium (141) nearly equaled the atomic number of uranium (235). She suspected that the neutrons they had shot into the uranium had split some of the large atoms into smaller elements. She was correct: The experiment had caused fission. Not only could one thing change into another in nature, but human intervention could cause this to happen. Three years later, under the bleachers of the Amos Alonzo Stagg football field at the University of Chicago, Enrico Fermi laid it all to rest. Shoot in one neutron and U235 goes up to U236 for instant and then splits, leaving smaller atoms and spraying out a few extra neutrons. These neutrons hit other uranium atoms, causing a chain reaction. Not only could you turn uranium into a different element, but, as Einstein predicted, you could turn matter into energy. The rest, as they say, is history.

The impact of Darwin's claim for a detailed operational mechanism of transmutation of one species into another seemed to the 19th-century mind to be just as difficult and unsettling to accept as Rutherford's claim that one element could change into a different one. How can one thing change into another?

Owen's View of the Problem

Sir Richard Owen brought this issue into sharp focus, as he studied and described the fossil record in the context of the rapidly advancing field of geology. There was a clear consensus that many of the geological strata (layers stacked one upon another) were deposited sediments (sand and clay made of tiny fragments of quartz and other rocks) that had subsequently lithified into stone (shale, limestone, sandstone). Layers of crystallized igneous rocks (basalt, quartz, and obsidian) had also formed from cooling magma spewed from a volcano, and could be found both above and below the sedimentary layers. Deeper in the layers, metamorphic rocks converted from the other types by heat and pressure (slate, marble, garnet) would appear. Charles Lyell had helped clarify all of this in the 1830s.

In these geological strata, a series of different horse species had been identified that seemed to show a natural progression of increasing size and decreasing toe number toward the large, modern type of horse that walks on a hoof based on a single toe. The different species appeared in discrete geological strata. In higher strata, the species were closer in appearance to the modern horse. Furthermore, a species found in one stratum was never found in any other stratum. It appeared that one species replaced another in a time-based sequence and that there was change in anatomy between steps. Owen's solution was that this progression of anatomic change within a species represented serial creation events by God.

We can smile as we consider Owen's philosophical contortions to achieve this interpretation. However, Owen placed great emphasis on a particular confounding factor after reading Darwin's *Origin of the Species*. In alchemy or nuclear physics, we expect to see a sudden abrupt transition from one element to another. However, in Darwinian Theory, there is an expectation of gradual transition. Shouldn't the fossils of species members found in the lower part of a stratum resemble more closely those members of the older species in the upper part of the stratum below? That is, shouldn't there be a gradual transformation from one to the next rather than a stack of discretely different species?

Rate of Change in the Fossil Record

Darwin insisted that the transitions were gradual, and that it was only the incompleteness of the fossil record that gave the appearance of sudden transitions. He did this in part to help avoid Owen's interpretation being applied to the data he was analyzing.

It seems quite clear now that both types of change—gradual and abrupt—do take place, although both are entirely consistent with Evolutionary Theory. The pattern of abrupt change is the essence of Stephen Gould's *punctuated equilibrium* theory (which states that the anatomy of a species may remain unchanged for long periods of time) discussed in Chapter 1 (see Eldredge and Gould 1972). It is also an important basis for Steven Stanley's range of ideas about "macroevolutionary" patterns (Stanley 1998) and Gaylord Simpson's older views about how "radiations" of large numbers of new species appear rapidly in the fossil record (Simpson 1944, 1953). It seems that often, a new species undergoes rapid change near its time of origin, but then changes very little over the following millions of years.

Under pure Darwinian analysis, we might expect species to always undergo steady accumulation of changes across time. If a species is in existence for millions of years, then the species members from the later period should look quite different from the members from an earlier period. In many cases, the fossil record doesn't seem to bear this out. In part, the problem is due to the methodology used by paleontologists (biologists who work primarily with fossils). When they discover new specimens from a later stratum that look different from the previously discovered specimens in the earlier stratum, they are apt to call the new specimens a new species. Thus, the methodology becomes self-fulfilling. If every time we find a difference we call it a new species, then there will never be any change in a species. If some species have relatively little change over long periods, this will be taken as evidence of *stasis* (no significant change over time). If other species change rapidly, then they are deemed a short-lived, separate species, each one with a short, static existence.

The Definition of Species

To understand what is taking place here, we have to consider the process of species formation. Again, the term for this is *speciation*—or, of course, *The Origin of Species*. Obviously, you can't get very far with explaining speciation until you have a definition of what a species actually is. Unfortunately, however, we don't really have a single, fully agreed-upon definition.

Ernst Mayr and the Biological Species Concept

The best accepted definition has been championed by Ernst Mayr, the Harvard University evolutionary biologist who played a major role in the development of mainstream evolutionary theory for more than 70 years. The last book Mayr wrote on the subject was published in 2004, a few months before his death; it was his 25th book, and he was 100 years old. As a former student of his, I could see in this book his attitude that he had to keep writing this again and again until everyone finally got it. This definition is the *biological species definition*. Simply stated, it asserts that two populations have split into two species when they can no longer successfully interbreed. There exists, therefore, a reproductive barrier to gene flow. A well-known example is the donkey (*Equus asinus*) and the horse (*Equus caballus*): A cross of a horse with a donkey will produce an infertile hybrid, a mule. If there is no reproductive barrier between two similar subpopulations, their genes will eventually mix through interbreeding until there is no clear difference.

Willi Hennig and the Phylogenetic Species

The most important alternate definition comes from the German entomologist (insect specialist) Willi Hennig, who wrote up most of his important ideas while being held as a prisoner of war by the British at the end of World War II. He was employed by the Germans (after he was wounded as an infantryman) and then by the British as a POW for the purpose of using his entomological skills for the prevention of malaria. What Hennig wrote out in longhand went through a series of revisions and publications, the most widely read being the 1966 English edition titled *Phylogenetic Systematics*. This small book has had tremendous impact in biology, inspiring nearly religious zeal in its followers. In large part this is due to the clean, logical principles that he explains with great economy and precision. Hennig's arguments in the book had an uncanny ability to identify problems in evolutionary biology that were due to imprecise reasoning, or to concepts whose consequences had not been fully considered to the last detail.

Hennig was concerned that, although Mayr's definition worked well enough in the here and now, there was no useful way to apply it across time, as in paleontology. We cannot know the breeding effects of two similar fossil groups in a given geological stratum, and, more importantly, we don't know whether a species member from the early years of a species existence can interbreed with a species member from a million or so years later. Hennig drew an example from the problem of classifying a fly. If we looked at the larval stage, we might think it was a kind of worm. Then, once it had undergone metamorphosis, we would have to reclassify the same individual organism as an insect. He points out that we must base our classification on a typical stage such as the adult form. This choice is based on the fact that we can discover the *phylogenetic relationships* (ancestry and relatedness to other groups) so that we know the fly belongs with other insects and not with the worms. This can be proven by comparing adults, fossils, amber trapped specimens, or, more recently, based on biochemical data. Hennig traces this fundamental understanding of how to approach classification to the ancient Egyptians by citing comments in the Ebers papyrus from Thebes written during the 17th dynasty period around 1550 B.C. (Hennig 1966). Hennig then takes the example of the fly and effectively applies it to the life of a species. Thus, the classification will be based on a single stage in the existence of a species. Every time a new species emerges from an old species, Hennig declares there are now two new species and that the parent species no longer exists.

Branch Points and the Origin of Species

Traditionally, it is thought that the existence of a species ends when it becomes extinct, leaving no survivors. In Hennig's scheme, a species disappears by giving birth to a new species. This is logically "clean" because we are unable to apply the reproductive barrier test to historical species from the fossil record. However, Mayr feels that this misrepresents the biology of the situation. Wheeler and Meier's 2000 book, *Species Concepts and Phylogenetic Theory: A Debate*, presents an excellent set of articles, responses, and rebuttals that cover many aspects of this debate. The followers of Hennig are often termed "cladists," and Mayr is convinced that they are so doctrinaire and dismissive of other views that they simply must not be reading what he writes: "I realize that it is apparently distasteful for a cladist to read anything not written by another cladist" (Wheeler and Meier 2000).

Although Mayr accepts that a reproductive barrier may arise across time, he finds the Hennigian position to be utterly unacceptable. A large population is existing in an environment, but a small group with a different variation becomes separated either by a geological barrier (floating to a new island) or by an ecological one (a desert splits a grassland). With the elapse of time, the small group incorporates a series of genetic changes and ultimately can no longer mate successfully with members of the original species. The small founder group is a new species. But what about the original group? Nothing has happened. It has the same range of variation, and there are no new major genetic changes. Somewhere in the distance on an island somewhere, their descendants have become a new species. Hennig suggests that on the day the small separated population becomes a species, the original population also becomes a different species than what it was before—even though the original population is not changed in any significant way.

In fact, Mayr's objection to Hennig's phylogenetic species concept is further expanded when you consider Stephen Gould's view that most changes happen fairly quickly during the founding event that initiates a new species. A larger parent species then tends to persist in a fairly static condition for long periods—perhaps having several new small populations split off to form new species before it finally terminates and becomes extinct. Despite its appearance of stasis, Hennig will assign this species a series of identities. Proponents of Hennig's view would counter that there is no such thing as true stasis, and that if time is passing then all the populations are changing. Reproductive isolation across time is a non-issue because, with the separation of two or three generations, time will provide a barrier to interbreeding. Therefore, periodically identifying a new species status is reasonable.

Mayr has always insisted that, without a biological definition, there can be no solution. Kevin de Queiroz, a research zoologist at the Smithsonian Institution in Washington, D.C., who has a passionate interest in cladistics, has argued that, in effect, Mayr didn't really understand the implications of his own claims (de Queiroz 2005). In the article, de Queiroz points out that there are more than a dozen different current proposals for how a species should be identified and that these proposals all produce different classification. Searching through Mayr's older writings, de Queiroz suggests some useful common grounds. However, reading the exchanges in the 2005 debate book (Wheeler and Meier 2000) leaves no doubt about what Mayr had to say.

Reproductive Isolation—Sufficient but Unnecessary?

Both sides agree that the triggering event for speciation is often the splitting off of a small subpopulation. Once this is accepted, then it becomes interesting to try to assess how small of a population is involved—is there a minimum size?—and how the groups remain separate. Although de Queiroz (2000) insists that we must abandon reproductive isolation as a requirement for a species definition, it is still the case that virtually all experts accept that if there is a definitive barrier, then a speciation event has indeed taken place. That is, all agree that reproductive isolation can be a mark of speciation, but not all agree that it must be present.

Mayr and other advocates of the Modern Synthesis, with its emphasis on gradual changes among populations, have always thought more or less the same way Darwin did on this question. If you separate two populations and allow enough time to elapse, they will become different enough genetically or in appearance so that the normal sexual reproduction system of gametes (sperm and eggs) and meiosis (production of the gametes) will be overwhelmed by the differences and thus begin to fail when any interbreeding is attempted. Once that happens, the separation of the two groups is irrevocable and speciation has certainly occurred.

Gene Spread in Species Large and Small

When a population is small, it is easy for a few members to have a large influence on the genetic makeup of the entire group. If there are just two males and two females in the new incipient species and three of them share an unusual genetic variant, it won't be long—one generation—before nearly all the individuals have the variant. By contrast, consider the human species: If a new useful genetic variant turns up in one new baby, how long will it take before that gene gets spread through the entire population of humans? If human population size stays around four billion, and we allow 20 years for each new generation, is it possible to estimate the number of years required for the new gene to become present in every living person?

If you relentlessly double the frequency with each generation (steady production of two children with the gene from each parent having the gene), it only takes about 35 generations (a mere 700 years). However, if the population remains stable, then people must be having only one child per parent.

Also, some descendants will develop negative mutations. The calculation gets quite complicated, but an estimate of up to several thousand generations is sometimes suggested. Recorded human civilization goes back just 250 generations. Current evidence suggests that the founder population for our species was as small as 1,000 individuals and that they lived about 60,000 years ago (3,000 generations). You can readily see that success in response to rapid environmental changes is going to favor species with small numbers of members. Not only can one new gene spread relatively rapidly, but several new genes can spread as well.

Chromosomal Speciation

In the 1980s, a new idea was proposed to explain the mechanics of speciation. This is a mechanism of genetics that can rapidly isolate two populations of a species. Even if the two groups are living among each other and look identical, the two may become separate species (as with Mayr's definition)—two populations unable to interbreed. Once this takes place, the gene pool of the two species will be separated from each other permanently and completely. Eventually, with the elapse of time and across many generations, the two isolated species will come to appear different from each other as well. The biology that underlies this is not so much in the genes as it is in the chromosomes. What is the difference between a gene and a chromosome? Well, a chromosome is actually a kind of package or "carrying case" that contains hundreds or even thousands of genes. The chromosome is composed of the DNA as well as a variety of proteins and RNAs that take part in reading, repairing, and copying the DNA and other proteins involved in folding and packaging the DNA.

A Primer on Chromosomes and Genes

Humans have 23 chromosome pairs. All of our many thousands of genes are packaged into these 23 pairs. They come in pairs because we all have two copies of each gene and two copies of each chromosome. You get one full set from your mother and one full set from your father. If you are male then you got a copy of your father's Y chromosome, and if you are female you got a copy of your father's X chromosome. Everyone gets an X chromosome from their mother. Aside from the X and the Y, the other 22 chromosome pairs are physical matches for each other. In the remainder of my discussion I will be referring to the shape "X" as a general description and not specifically in reference to the sex chromosome, XY issue.

There is nothing magical about the number 23 for chromosomes. Species can have 8, 21, 24, or 57 chromosome pairs; the exact number varies widely among species. However, virtually every individual member of a species has exactly the same number of chromosomes as all the other members of that species. What happens if an individual is born that has its DNA packaged into a different number of chromosomes? When that individual mates, the offspring will have bad chromosome pairing. This turns out to be a significant problem. However, because of some particular details, the chromosome issue has a means of producing smooth efficient speciation.

A Detailed Model of Rapid Speciation

Details of the replicating chromosome

Let's find out how chromosomal speciation can take place in a single individual organism in a single generation. Consider a typical chromosome that is shaped like an X and is made up from two V's stuck together at their tips (see Figure 4-1). In each X, the left side is one copy of the DNA and the right side is a second copy. The left and right sides are joined at the *centromere* to give an appearance of an X. The drawing starts with two chromosomes, one made up from two large V's and one made up from two small Vs. It's worthwhile to look at the figure and follow along closely, because the process of sudden speciation shown here helps explain a great deal of how evolution works, and probably shows how the human lineage split off from its ancestors (see Kehrer-Sawatzki 2005).

The chromosome images on the left are shown in their appearance in the gametocyte stage after part I of the two-step meiosis process (see Figure 4-2). Completion of the second step of meiosis, which splits our Xs down the middle vertically, produces the final genetic contents that will be packaged inside the gametes (egg and sperm).

The starting point

In species A, the adults will have two copies of the big X and two copies of the small X. As germ cells are formed to make the gametes in the sperm and egg, a process called meiosis takes place in which only one copy ends up in each gamete cell (A1 and A2), instead of the two copies in all the regular somatic cells in the body. In the first line of Figure 4-1, we see a mating between gametes from species member A1, the mother, and species member A2, the father. The result is an offspring (A3) that is the same as all the other members of species A at the chromosomal level. It has a pair of small X's and a pair of large X's in nearly every cell in its body. When it forms its gametes, they will have the same chromosome appearance as the gametes it received from its parents. This is what happens most of the time in a species.

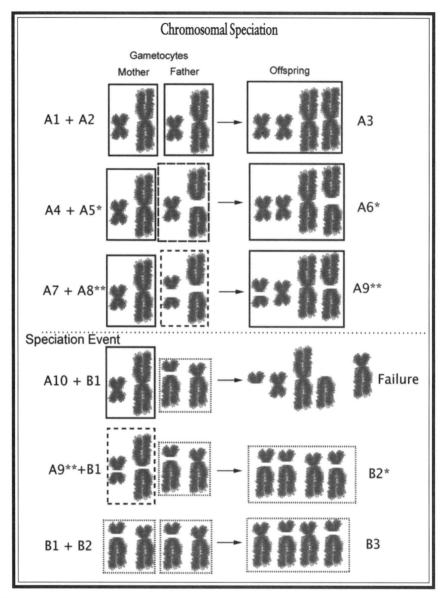

Figure 4-1: Mechanics of chromosomal speciation. A new daughter species (B) can originate suddenly when a new chromosome arrangement develops in the parent species (A). A fracture of a large chromosome occurs in the germ cell line in individual A5. Next we see two fissions in descendant A8. When a new type of chromosome develops from a fusion of the fragments in B1, the stage is set for a speciation event. These chromosomes are depicted as they appear in the gamete-forming cells in the metaphase I stage of meiosis (see Fig. 4-2).

A chromosome fractures and puts the species in play

In the second line we see what happens if member A5 doesn't contribute a normal large X. In this individual, the large X has physically broken into the two V's that are normally hooked together as the large X. This is a process called centromeric fission (see Perry 2005 for details). As it happens, there is "no harm, no foul" in this. When a gamete from A4 meets a gamete from A5, they are still able to produce an entirely normal offspring (A6). Fortunately, when the chromosomes line up, as they must eventually do in the process of forming new gametes, the two loose, big V's are able to line up opposite their counterparts that are still stuck together in the chromosome donated by the A4 mother. This is what is shown for A6. The offspring is fully a member of species A.

A second break opens the gate

In line three, we have a gamete from a father (A8) who is probably a descendant of A6, and so has the condition of two large V's instead of a large X in many of his gametes. However, in A8, a second break has happened. Now, the two little V's in the small X have also separated. Once again, there is no harm in this. An ordinary gamete from mother A7 joins with the gamete with the two split chromosomes from A8, and they produce a perfectly normal offspring (A9), that is normal in appearance, fully fertile, and fully a member of species A.

A chromosomal fusion is the moment of foundation

I have labeled A9 with two asterisks because it will be a bridge across a speciation event, thus making what is seemingly impossible turn into a reality. This is because one of A9's siblings or offsprings has generated a new type of gamete (B1), that is about to become a new species with only a single individual in it. This is what seems to be impossible. If a new species is defined by reproductive isolation, how does it all get started? If there is only one member of the new species, who can it mate with? Look closely at what happens on line 4 of the illustration.

When you compare the makeup of the B1 gamete to the makeup of the A8 gamete, you'll see the problem. In B1, one of the small Vs has become attached to one of the large V's, making a new kind of X that we might call short-and-long. When a gamete from B1 is assembled with a gamete from A10, there is a full set of genetic material. However, as in the mule, when it comes time to form the gametes by lining up all the chromosome pairs and performing meiosis, disaster strikes—the match-up cannot be accomplished. In 999 times out of 1,000, the resulting gametes have incomplete or disordered gene sets, and the offspring will be sterile like a mule. Now we can see what exactly makes up a barrier to reproduction!

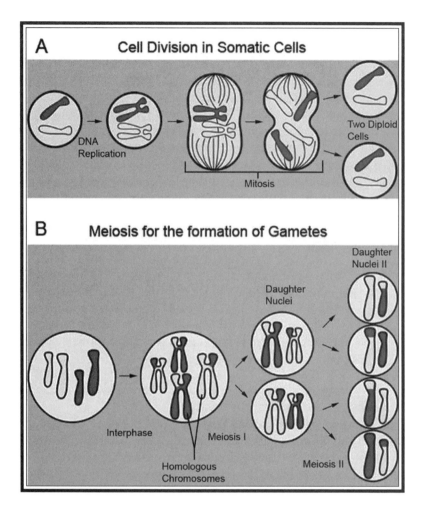

Figure 4-2: Chromosomes in mitosis and meiosis.

A) In standard cell division throughout the body (*mitosis*), each new cell receives an identical copy of the genetic information. A chromosome is a tightly coiled package containing a very long DNA molecule wrapped up in chromatin proteins. Each chromosome is present in two similar copies in each cell.

B) The process of making gametes (DNA in sperm and egg) is different because it (*meiosis*) allows for the creation of unique new chromosomes, and because the gametes have only a single copy of the chromosome. As in mitosis, the DNA is first replicated, and the sister chromatids are joined at their centromere to resemble an X. However, now the homologous pair members must find each other and line up (this is the process that is disrupted in a chromosomal speciation event). There are then two rounds of splitting, meiosis I and meiosis II, to produce the gamete.

A new species is reproducing

Now look at the fifth line to see how B1 manages to bring forth the new species and avoid immediate extinction. He has to be fortunate enough to mate with A9 or one of her relatives with a similar chromosomal structure. If that happens, the offspring will be a normal-appearing individual with a full genetic complement and capable of reproduction. However, the offspring, (B2) will be the first individual in the new species. What happens next to keep the species going? Species B needs a few generations of crosses with descendants of B2 or A9. Once there are a few individuals, some will end up with a pair of short-and-long X chromosomes. These individuals will be able to mate successfully with any B species member, and will now be infertile with the vast majority of A species members. With time, the floating V's in each of the species tend to reconnect as X's. Once that happens, the genetic bridge between the two species is closed and the speciation event is complete.

The gate closes, isolating the two species

Most chromosomes in most species are seen in the X pattern rather than the V pattern, although both types occur. This reflects a tendency of chromosomes to form into X's, so we have all we need for a definitive chromosomal speciation process. The split in an individual is an occasional anomaly that has the sole effect of opening the door to a possible speciation event. The development of a second split before the first one closes further increases the opportunity for an event such as the one shown in Figure 4-1 and described previously. As the "floating V's" proceed to reconnect back into their usual X arrangement, the bridge closes. If no new species arises, then this is inconsequential. If a new species is initiated, however, then the resumption of normal X configuration closes the door and completes a definitive and complete speciation process, with no possible forward progression of gene flow. The occurrence of any more classical Darwinian speciation pressure, such as environmental or geographic separation, during one of these chromosomal events will only accelerate the separation in advance of disappearance of the "floating V's." That is the "how" of the origin of species.

The Impact of a Biological Mode of Speciation

The behavior of chromosomes in normal cell division was described by the German anatomist and physician Walther Flemming in 1882, although he did not know the role of chromosomes in heredity. The behavior of chromosomes in meiosis for gamete formation was described in publications by German biologist and theoretician August Weismann in 1890, and by Thomas Hunt Morgan in 1911. It was Morgan—an American geneticist who won the Nobel Prize for his discoveries in his Columbia University laboratory—who was finally in a position to understand the significance of the entire process. Together with the rediscovery of Mendel's work in 1900, these discoveries ultimately led to the full understanding of the role of chromosomes in inheritance. Of course, Darwin could not have had any knowledge of all this. Those who are responsible for Modern Synthesis of Evolutionary Theory, however, were very well aware of chromosomes, but have tended to dismiss or minimize chromosomal speciation for methodological, theoretical, and tactical reasons. The important thing about this crash course in biology is that it shows that speciation—the formation of new species—is not necessarily due to mutation, adaptation, or physical isolation. Yet it has the power to create a new species abruptly, starting out with just a few individuals. Further, it may trap a small subset of genes to create a new type of animal, even if the new kind of animal is not as well-adapted as the individuals of the parent species.

If the new species has a particularly bad genetic makeup or has very bad luck in its environment, it may go extinct. However, if all goes well, the classic Darwinian mechanisms of adaptation to the environment and survival of the fittest begin to act upon the new species, gradually "tuning" and optimizing the new pool of alleles into a progressively better-adapted species. Yet another extreme interpretation, however, is that, in some cases, the speciation event captures or traps one single critical mutation that effectively defines the new anatomy and ecological niche of the new species. In this case, the process of evolutionary change, just like the process of speciation, is completed in an instant of *chromosomic centric fusion* (the reconnection of the two Vs). This is more or less what may have happened with emergence of the original member of the vertebrate lineage. Chromosomal speciation capturing the inverted embryo mutation would cause appearance of the new phylum of the vertebrates more or less in a single generation.

A Biological Basis for Understanding Form and Family: Homology and Systematics

Introduction

John Hunter, the great surgeon and anatomist of 18th-century England, died suddenly in 1793 at the age of 65 after storming out of a meeting about admitting medical students at St. George's Hospital in London. He left behind a personal collection he had assembled of 13,000 preserved specimens covering a vast array of materials from human anatomy, medicine, and embryology, including birth defects and "monstrosities." In addition, there were numerous fossils from around the world, as well as a most remarkable collection of comparative anatomy, including detailed dissections of tissues and organs from a wide variety of species of plants and animals.

Aside from the unfortunate impact of John Hunter's death, this "Hunterian" collection suffered two major unfortunate events. Hunter's brother-in-law, Sir Everard Home, who succeeded Hunter as chief surgeon at St. George's, apparently systematically burned most of Hunter's records of the collection so that he could hide his own plagiarism. (Home was trying to publish reports on the collection as his own.) In 1941, a German incendiary bomb struck the London museum housing the collection at the Royal College of Surgeons at Lincoln's Inn Fields, destroying thousands of specimens. Between these two events, however, the anatomical details revealed in Hunter's specimens played a fundamental role in the emergence and development of Evolutionary Theory.

Shortly after Hunter's death, his famous collection was purchased by the government and given as a bequest to be kept at the offices of the Royal College of Surgeons, newly completed in 1797. Following the unfortunate stewardship of the collection by Everard Home, one of Hunter's former surgical students, John Abernathy (then-president of the Royal College of Surgeons), took steps to correct the situation. In 1826, he brought in a young surgeon, 22-year-old Richard Owen, with a plan to set up the young man with a private practice at the Lincoln's Inn Fields. While developing his practice, Owen would be supported by a job at the Royal College requiring him to work from scratch, and examine and recatalog the entire Hunterian collection of 13,000 items.

The young Richard Owen completed the task in four years, and was intellectually transformed by the project. Among his other talents, he had learned French, so he was asked to serve as a tour guide when the world's leading biologist, Baron George Cuvier, arrived from France in 1830 to see the collections. Cuvier was impressed with Owen and invited him to spend some time looking at the collections of the Natural History Museum in Paris. Owen's visit to France took place a few months after the Great Academy Debate between Cuvier and Étienne Geoffroy Saint-Hilaire. Nonetheless he was exposed directly to the matters that had been debated, as well as to the intense emotions, rivalries, and differences in views of nature that underscored the debate.

It is clear from Owen's subsequent writings that he initially sided with his host, Cuvier. In 1833, Owen had begun writing up materials collected by the young Charles Darwin (then 24 years old) during the voyage of the *Beagle*. There is clear evidence of his support for Cuvier's "functional" view (in which the form of a structure is optimized or adapted for its function) of anatomical features in Owen's writing at that time. He may have influenced Darwin by his thinking. However, the bequest of the collection included a requirement that the college organize a series of public Hunterian Lectures, and Owen was assigned to prepare them. Owen's inaugural lecture in 1837 was attended by royalty and by Charles Darwin. The handwritten lecture notes for a critical lecture in the series have only recently been rediscovered (Padian 1995). They make clear that Owen had, by that year, turned around 180 degrees in his thinking and was coming to view biology from the point of view of Geoffroy Saint-Hilaire, who held that intrinsic aspects of the form of the animal were often more important than the function of a given structure.

Floating in the throat of every person is a small bone called the hyoid. It is shaped like a boomerang, and it does not actually contact any other bone directly. It plays some role in swallowing and it hangs from the tips of two odd little spikes that project from base of our skulls. Many muscles in the neck and the tongue connect to it. In his recently rediscovered 1837 lecture, Owen writes that the hyoid is related to an arch of the hypothetical third vertebra of the head. At the same time he describes it as being related to the third gill arch in fish. John Hunter had evidently been fascinated by the gill arches, as well as by the hyoid. He had made very detailed anatomical preparations of the gill arches in fish and amphibians. In a separate category there were a series of preparations of mammals to show the hyoid as part of a vocal system. By 1833, Owen himself had made several new preparations of the throat and hyoid in various animals to add to the Hunterian collection.

I believe that this anatomical conclusion—relating an odd, isolated, free-standing bone in the vocal apparatus of the mammals to one of a series of gill arch structures in fish and amphibians as well as to a part of a vertebra—was the defining intellectual event in Owen's career. His inspiration is clear. At the March 22, 1830, debate in Paris, Geoffroy presented this idea, which was published preliminarily in 1830 and then in full detail in 1832. This idea seems to indicate that there is a fixed set of items, real or metaphysical, available for animal construction, items that can then be re-formed into widely different shapes and functions among all animals. Nature radically reshapes and reassigns the usage of one shared structural element, rather than simply generating a new structure for a new task. With a bit of anatomical work, these identifiable structures can be located and their transformations described. The existence of such a thing in the animal kingdom clearly had a tremendous impact on both Geoffroy and Owen. Cuvier felt that the gill arches and the bits of the hyoid in various animals were entirely independent bones that existed for independent reasons in the various species; to him, this whole "continuity" concept was just so much nonsense.

Here again, Geoffroy was correct regarding a fundamental aspect of animal design. In the case of the hyoid, it is now possible to say with great certainty that this odd little bone is indeed related to the very gill arch in fishes, as Geoffroy and then Owen stated (see Figure 5-1). We can now say with great certainty that it is populated during its formation by neural crest cells arising near the sixth *rhombomere* (brain-forming segment). Further, we can now say with absolute certainty that the neural crest cells that help form the hyoid are unique relative to their neighboring cells, in that they express the *Hoxb-2* and *Hoxa-2* genes. Individual features along the dorsal to ventral portions of the bone are responsive to *Dlx* genes *1* and *2*, then *5* and *6*, then *3* and *7* (Kuratani 2005a).

In this way, parts of this little bone can be uniquely and specifically related to certain structures in a wide variety of different vertebrate animals because they share specific morphogenetic gene expression properties. For instance, if we can locate a few cells that express products controlled by the eight genes mentioned previously, then we can tell what the "address" is in the body for those cells. This is without looking at their appearance or the tissue they were taken from. Without ever seeing the fully formed structure, we know that in virtually any vertebrate, they will normally develop to be either part of a hyoid bone or a second arch gill structure. The point of this bit of detail is to immediately debunk at the outset of this chapter a pernicious idea that needlessly troubles some biologists—namely, the idea that any statement about homology as a reflection of the "true" origin of a particular bone is rooted in mysticism and/or ancient Greek platonic idealism. It is all well and good for a modern-minded biologist to reject ideas that are rooted in ancient philosophy. However, time moves on and the situation has changed.

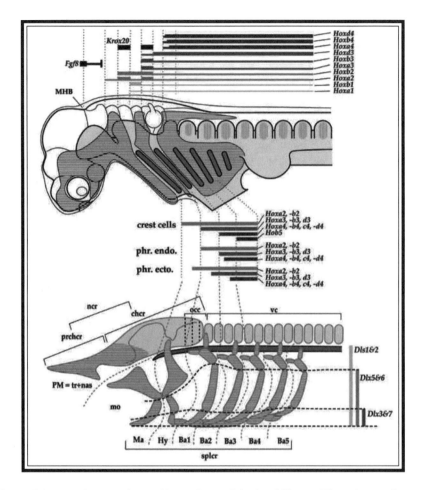

Figure 5-1: Morphogenetics and homology of the hyoid bone. The advent of modern morphogenetics has made it possible to reveal genetic positional markers that provide a fixed identity for each of the various skull components. The hyoid bone (Hy) played a key roll in the development of the ideas of unity of body plan (Geoffroy) and homology (Owen). It is specified by *Hox a2*, *Hox b2*, *Dlx 1* and *2*, *Dlx 5* and *6*, and *Dlx 3* and *7* genes, as shown where the grid lines intersect.

There exists an incorrect assumption about scientists, such as Etienne Geoffroy Saint-Hilaire or Sir Richard Owen, who worked in the 19th century—namely, that what they observed is useless given our modern knowledge. In fact, it is true that no matter how far-sighted they were, those older scientists could not possibly have used the same language we do, and could not have had access to the detail and theory elaborated in the decades that have elapsed since their work. But these were brilliant people who examined astonishing amounts of data, and who accurately described patterns that emerged from the background of the detail that they witnessed. As I have already pointed out, homology is a bitterly contested concept that has been critical to much of biology. It has a history that is worth recalling, but its meaning is being redefined in the present. It underpins the study of biological form (or *morphology*, as Goethe originally called this area of research); it is essential to being able to recognize which organisms are related to which other organisms; and it is essential to understanding how evolution changes organisms as they adapt under natural selection.

Fortunately, the clouds of uncertainty that hung over this critical subject are now starting to clear. It is true that a great mass of intellectual effort over the past 130 years has gone into an effort to banish homology, because of its roots in Owen's mystical archetype. However, homology has been rescued by molecular biology. It is definitively here to stay, and efforts to do without it are now probably pointless as well as misguided.

Homology

Although the morphological genetics of the *Hox* gene system is in the process of providing a real basis in biology for similarity among animals at its most basic level, the historical aspects of the use of the concept are still helpful in attaining a full understanding of the issues. Since Sir Richard Owen provided formal definitions of for three kinds of homology (fundamental similarities in anatomical structures) in the 1840s, his view has served as the basis for much of modern organismal biology.

Geoffroy Saint-Hilaire and the Biological Roots of Homology

Before Owen, Étienne Geoffroy Saint-Hilaire had suggested that there was a fundamental set of body components that appeared again and again in all of the species that fit into his "unity of design" concept. He pointed out that if he compared embryos of two groups, he could often see similar structures that were difficult to identify as similar in adults. One animal might have two bones serving as part of the skull, whereas in another, the two bones would fuse together and separate from the skull to form part of the shoulder. Even though they had a different appearance and different functions, Geoffroy showed that they shared a fundamental identity at some level. Geoffroy identified bones by the existence of an independent center of ossification (bone formation) in the embryo, and by a consistent series of connections to tissues around it. These two principles were important in that they identified a structure by its relative position in the animal, with no reference at all to its shape or function in the adult. This allowed a given structure to be remodeled into different adult forms and put to different uses, yet still be essentially the same thing.

Geoffroy did not provide our definition of homology. Although it can be argued that his idea represented platonic idealism, Geoffroy did not have a philosophical agenda to support Platonism. Rather, he was describing a biological phenomenon, and virtually all modern biologists will agree that he correctly interpreted what he saw. In fact, because he related the similarity among structures to their position in the embryo and their surrounding connections, his approach has proven to be extremely close to our modern morphogenetic method of determining the genetic identities of various tissues.

Sir Richard Owen Defines Homology

Sir Richard Owen formalized this sort of similarity as *homology*, which he defined initially in 1843, and then in full detail in 1848. He could not really explain why this phenomenon occurred. He related it to his model of an archetypal form from which all other vertebrates were made by reshaping of parts (see Figure 1-2). To avoid the metaphysical issues, he insisted that he was just trying to simplify anatomical nomenclature. There was no reason to have different names for the upper arm bone of a cat and a person; both could be called the humerus, and were so similar in shape, function, and connections that there must be some true, underlying identity linking the two. He called this *homology*, a term that is still used in this way today.

Although it is essential to a wide variety of biological disciplines, the term homology is problematic in that it is so "theory laden" that is it is difficult to say what it means outside of a given theoretical context. Even today, there are many different theoretical contexts applied to the term. It may be helpful to point out that Owen defined three different types of homology. *General homology* was meant to describe the relationship of any body part to the corresponding part in his archetype. *Special homology* (the word *special* refers to species) is his best remembered definition: "the same organ under every variety of form and function" when compared between two different species. Finally, *serial homology* refers to structures that share an identity from one segment to the next, such as the spinous process on each vertebra or the legs on an insect. When two structures appear similar but have different fundamental origins, we say they are *analogous* rather than homologous. The classic examples of this are the wings of a fly and the wings of a bird. At the time of Owen's publication, these terms yielded tremendous clarification in biology.

In order to help complete the array of terms, Hennig has emphasized *homoplasy*—the development of the same condition independently in two related species. An example is the character state (the genetically based feature of an organism's anatomy or biochemistry) of tail loss that is observed in hominoids (apes and humans) and in the Barbary macaque (*Macaca sylvana*). These two taxa have the same feature—in this case a negative one, tail loss—but it seems clear from a wide variety of evidence that this happened quite independently in the two groups and thus has no value for suggesting any special relationship between this species of macaque monkey and the lineage of the apes and humans.

A Rebirth of Owen's General Homology

Sir Richard Owen's definition of general homology had a very limited theoretical context, and was widely dismissed as soon as the archetype concept was abandoned by most biologists after the publication of Darwin's *Origin of Species* in 1859. However, as we explore the finer details of *Hox* genetics, there will be a need for a new term that expresses this concept. This will describe the relationship of a particular structure to a given aspect of the segmental assembly instruction. For Owen it was a search to find, for example, the shoulder blade in a part of his theoretical ideal vertebra. Currently, however, we look for genes that produce different structures in different segments—as with the *Antennapedia* gene that revealed what Owen would have called a general homology between both antenna and leg on the one hand, and the antennapedia gene product on the other.

Alternatives to Owen's Special Homology

Special homology as it was defined by Owen is hard at work today throughout biology. It is applied to proteins, DNA sequences, cell types, genes, organs, and a wide variety of anatomical features. However, there has been a lingering discomfort about what it really means (Muller 2003, Fitch 2000). This lack of clarity is particularly problematic for cladistics. As much as Hennig wanted to move away from reliance on morphology, it is still the case that much of the work on classifying animals—and virtually all of the work on classifying fossils—relies on identifying a feature that is homologous between three species, and showing how it has shared a change in two of them. If the change is a homoplasy (originating independently twice), the cladistic conclusion will be incorrect. However, how can we assure homology if we don't know the relationships among the species to start with?

Serial Homology and Hox Genetics

Despite all the work over many decades involving Owen's special homology, serial homology has languished. In part, this is because of some negative fallout from the fact that William Bateson paid an inordinate amount of attention to serial homology (see Figure 5-2) near the turn of the previous century. Bateson played an enormously critical role in the history of science, but his legacy remains unsettled (Bateson 2002).

Figure 5-2: Major figures in biology and evolution with key contributions. Carolus Linnaeus (1707–1778), classification and naming. Johann Wolfgang Goethe (1749–1832), modular theory of similarity among organisms. Étienne Geoffroy Saint-Hilaire (1772–1844), unity of plan in vertebrates and invertebrates. Sir Richard Owen (1804–92), homology. Charles Darwin (1809–82), natural selection, common descent. Gregor Mendel (1822–84), pattern of genetic inheritance. William Bateson (1861–1926), mutations, homeotics. Willi Hennig (1913–76), cladistics classification. Ernst Mayr (1904–2005), 20th-century Evolutionary Theory.

Bateson, Mutationism, and the Discontinuity of Evolution

As a young man, William Bateson worked with Darwin, but eventually he began to question Darwin's gradualism. He felt that evolution of morphologies tended to show sharp discontinuities rather than slow, continuous changes. Using the best methods available in the 1880s, he collected information on hundreds of variations and mutations from throughout nature. Many of these concerned the fascinating aspects of variants in serially repeating structures that he called *meristic* or *homeotic* changes (variations related to the position of a segment relative to other segments in the body). These were categorized, summarized, and interpreted in his major work published in 1894 by Cambridge University, titled *Materials for the Study of Variation Treated with Especial Regard to Discontinuity in Origin of Species*. In the book, Bateson concludes that variation and inheritance must be based on discontinuous changes. This is also considered to be the foundation work for the theory of *mutationism*, in which inheritable mutations play a major role in generating new kinds of species. Given this fact, Bateson believed that Darwinian natural selection acting on small gradual variants could not explain the origin of species in most cases.

Mendel's Discovery

Although the other scientists of his era were unaware of his work, Gregor Mendel discovered in the 1850s that inherited variations were discontinuous. When you breed a red flower and a white flower, the result is red or white, but not pink. This has to do with how a red gene and a white gene are sorted during the reproduction of a flower. There was no evidence in Mendel's experiments that any blending normally took place.

Mendel demonstrated the way in which two alleles for a gene will show themselves in the offspring of parents that possess two different versions of the gene (two different alleles). For instance, if a single gene controlled the color of a flower, then there might be a red allele and a white allele. One allele might be dominant over the other. Biologically, this might mean that the flower began its development as white, but if the gene became active then a red color appeared. If the gene remained inactive, the white flower color persisted as the plant grew. If both alleles—the two versions of the gene from the two parent plants—were inactive, the flower would be white. If either parent passed on the red gene, the plant would have one red version of the gene and one white version, and would produce a red flower. If both parents passed on a red version, the two red genes would produce the same red. Overall, if each parent had one red and one white allele, it turns out that there would always be an average of one-quarter of the offspring with white flowers, and three-quarters of the offspring with red flowers. In shorthand, if a capital R represents the dominant red gene, and a small r represents the recessive version producing a white flower, any plant could be described as R/R, R/r, or r/r, depending on which pair of alleles it received.

These simple inheritance patterns observed by the monk Gregor Mendel in the mid-1800s provided the groundwork for genetics, because they showed that features on a living organism could be tracked as if they came from specific, traceable locations in the organism's inherited definition. These locations are the genes that we now understand in great detail.

The Rediscovery of Mendel's work

In 1900, three plant varietal researchers in Europe—Hugo de Vries, Carl Correns, and Erich von Tschermak—independently came across Gregor Mendel's 1866 paper on plant hybridization and cited it. Correns had found Mendel's work during the course of his research, but did not initially appreciate the full significance. De Vries was shown a copy of Mendel's paper by a professor reviewing the de Vries report prior to publication. De Vries responded by changing some of his terminology to match the older paper, but initially decided that it was not worth citing. By this time, Correns had appreciated the some of the importance of what Mendel had done and noticed the evidence that de Vries had read Mendel as well. By the end of 1900, three separate publications appeared with citations to Mendel. Each scientist was publishing his own important new observations on hybridization, but now found his work anticipated in an obscure paper by a monk written 34 years before. The story is told that Bateson, reading the de Vries paper on the train ride from Cambridge to London, understood abruptly in a moment of brilliance that the fundamental problem of inheritance had been solved by Mendel. The discontinuous basis of mutational change that he had been searching for had already been found.

Bateson translated Mendel's work into English and became its most outspoken advocate (Bateson 1902). He coined the term *genetics* to describe the new field he launched that day. Oddly enough, Bateson's apparently undeserved reputation for being wrong about important things seems to stem from the bitter attack he faced by those academics who opposed the idea of Mendelian particulate genetics. In particular, two leading Cambridge academics, Karl Pearson and Walter F.R. Weldon, engaged in a protracted intellectual assault that advocated the gradual character shifts described by the discipline of biometrics, as opposed to discrete character conditions carried on genes— which they dismissed as an odd, special case in the peas that Mendel had worked with.

Later, as evidence that genes resided on chromosomes began to accumulate, Bateson was slow to accept this additional aspect of genetics. This left

Bateson with three strikes against him: He opposed Darwin's gradualism, he fought endlessly with the respected biometricians, and he was slow to accept the importance of chromosomes. His intense focus on variations in repeating structures—Owen's serial homology and his own meristic variation—was little understood, drew little attention, and seemed tangential.

Fisher and population genetics

Ronald Fisher rescued Darwinian gradualism in the 1930s by showing that even with discrete Mendelian genes, if population mathematics were applied, it all looked gradual, as Darwin had said (Fisher 1930). Therefore, aside from Bateson's other "sins," his efforts in repeating structures seemed fruitless—they did not disprove Darwin's fundamental view, they did not prove to be the road to understanding inherited change (Mendel's work took over that prize), and they made no sense from the point of view of chromosomal or population genetics. After 1900, this work wasn't really resumed with great effect by anyone for nearly 80 years.

In addition to coining the term *genetics*, William Batson also invented the term *homeotic change*. As has Geoffroy, Bateson has now been extensively rehabilitated, and he is now considered the pioneer of the field of homeotic change or *Hox* genetics (Carroll et al 2005).

Flower and the Discovery of Field Homology

There has been mounting evidence since Owen's time that his three types of homology—general, special, and serial—may produce the same result for a given feature, but that occasionally, a species is described in which an odd exception to this occurs. A structure with a unified identity in all three approaches may undergo what can be called a *homological transmutation*, in which an aspect of that feature literally changes its identity. Similar to atomic nuclei and to species, a given structure may be able to literally turn into something else. Biologists who are happy to accept transmutation of species are often not so happy to accept transmutation of homologies. It is the same issue covered in the introduction of the previous chapter; How can one thing change into another? What sort of chaos will result—biological or intellectual—if we accept the occurrence of this kind of change? We now understand that this can occur because multiple morphogenetic influences can act on a given embryological tissue, and the balance and relationship between these influences can change with mutation. A major step in evolution of upright posture in humans appears to have involved one of these changes in the spine, one that can be called a *homological discontinuity*. It has only recently become possible to understand how this could have happened (Filler 1986, 2007a).

Working around the same time as Bateson, William Flower also became involved in Owen's serial homology problem. Flower was a military surgeon who served in the Crimean War in the 1850s, and then became curator of the Hunterian collections at the Royal College of Surgeons in London (a position Richard Owen had recently relinquished after 25 years). He published *Osteology of the Mammalia* (a thorough description of the bony anatomy of a wide variety of mammal groups) in 1876, and later, when Sir Richard Owen died, Flower again took over Owen's old post by becoming director of the Natural History Museum in London.

Flower was in communication with William Bateson, and would have been well aware of Sir Richard Owen's ideas. In my own studies of serial homology at Harvard in the 1980s, I realized that Flower had stumbled onto a serious flaw in Owen's concept of homology. The issue may have seemed minor and limited when Flower encountered the problem; however, it now appears to be of growing importance for two reasons. Firstly, it goes to the heart of the issue of genetic control of embryological development. Secondly, it helps make sense of a critical event in human origins.

Failure of the homology paradigm

The problem, put simply, is that biological mechanisms that moderate serial homology can produce a conflict with the rules of special homology. That is, a structure that is homologous to one structure from one point of view can be proven to be homologous to an entirely different structure when evaluated from the point of view of serial homology. The solution has been to propose a controversial new type of homology (de Queiroz 1985) that is capable of overriding classical homology. As I point out in my 1986 Ph.D. thesis, the problem is overwhelming. There are enough critical examples to prove that there is not really any such truly fixed thing as homology. It is an epiphenomenon, or an external appearance, that results from peculiarities of embryonic development.

A variety of scientists have attacked homology as a concept because it traces to Owen and appears to rely at is most fundamental level on platonic idealism at best, and mysticism at worst. Nonetheless, it has been feared that the loss of homology as a fixed biological fundamental would cause chaos in a number of subfields. Chaos or not, homologies can change, and this possibility must be incorporated into any theoretical structure that depends upon homology. Granted, it happens only occasionally—typically where serially repeating structures such as the axial morphologic modules are involved. Nonetheless, these homological shifts or discontinuities must be allowed for in any comprehensive model of the genetic basis of homology.

Cladist Analysis of Lineage Relationships

If we compare the hyoid bones of three different kinds of animal to one another, there is a reasonable chance that whichever two kinds of animals have the most similar hyoid will be the two that are most closely related. This general concept applies to any bone or protein or other feature we might select. Looking at modern animals, we can rely heavily on protein and DNA to determine which species are more closely related on one another. If we are going to include fossils in the analysis, however, it is helpful if we can make the judgments about relatedness based on the features or shapes of bones.

The Project of Naming and Classifying

It is fascinating and informative to find out that whales and dolphins are most closely related to the hippopotamus among all other living mammals. The anatomy of the sea mammals is so different from that of the hippopotamus that this was difficult to verify with certainty until modern biochemical methods for comparing homologous proteins became available (Gingerich 2005). The ancestors of these modern sea mammals emerged out of the group we call artiodactyl ungulates (the cloven-hoofed, plant-eating animals such as the gazelle and the giraffe along with the hippopotamus). This kind of information helps us know how to recognize key changes in fossils, and it helps us appreciate how evolution works. It is also the central subject of a related field called *systematics*, the work of naming and classifying species into appropriate groups. This was the original project of Carol Linnaeus in the 1700s. He named and classified tens of thousands of different species into categories of genus, family, order, class, and kingdom.

The ungulate species are animals that are grass-eaters and walk essentially on their toenails. This includes the order of perissodactyls—horse, tapir, and rhinoceros, which have only one toe (or a hand/foot centered on the third digit) reaching the ground at the end of their legs—and the related but separate order of artiodactyls—antelope, cow, goat, camel, giraffe, and hippopotamus, which walk on hooves composed of two toes. Because of domestication, ungulate species, artiodactyls in particular, are of great interest.

The order Cetacea (whales and dolphins) includes many sharp-toothed carnivores, so many biologists expected to find that cetaceans were descended from a type of carnivorous mammal. In the 1880s, William Flower proposed the link between cetaceans and artiodactyls, relying in part on the unusual aspects of the serial homology of the vertebrae. Later morphologists emphasized the teeth and skull, and applied a method formalized by Willi Hennig called cladistics to analyze the information from the bones to show Flower was wrong. Next came the biochemical information, again linking cetaceans and artiodactyls. The cladistics specialist responded by poking holes in the variable conclusions that could be generated from biochemical analysis on the subject (Wys et al 1987). Finally, when ankle bones of the most ancient known cetacean group were found, the morphology strongly confirmed Flower's original opinion and settled the position and history of cetacean origins.

The type and quality of the data examined clearly has a great effect on the conclusions no matter what the method of analysis. This was a major example of applying cladistics methodology to try to confidently refute both the "old" conceptual morphology of William Flower, as well as the new data from biochemistry. The cladistics method of analysis is important, though, and very often produces results that match closely with the biochemistry.

The Basics of Hennig's Cladistic Method

Hennig very carefully describes his method (Hennig, 1966) although it is not always properly understood or applied (see Figure 5-3i). We start with a stem species (A) that has a feature or character a. It may have a descendant species (B) with the same character a, but when it has a descendant species (C) in which a is changed to a', then Hennig gets interested. We now have a derived or changed character that Hennig calls an *apomorphy*. When C then undergoes speciation, we will have new species D and E, and they both will carry a' or some further change a". Because both D and E share the derived character they inherited from C, we say they have a *synapomorphy*, or shared derived feature. Now if a biologist discovers living species B, D, and E, but has no fossils of extinct A or C to work from, he can still sort out the lineages as long as he can tell that a' is the modified or altered version of a. He will find that D and E are sister groups because they have a shared derived character. The group of species that includes C, D, and E is a *monophyletic* group (having a single ancestor), at least from the time when a' first appeared in the C species before it was split by speciation.

Species B has the old form of the character a, as does the extinct species A. Hennig calls the old form a *plesiomorphy*, and the fact that both A and B have the old form is a *symplesiomorphy* (shared ancestral character). In Hennig's scheme, symplesiomorphy is never of any use for working out phylogenetic relationships.

Features of a Cladogram

Another feature of Hennig's evolutionary lineage trees that differ from some more traditional depictions is that in his tree *cladograms*, all the branches have exactly the same length. In other systems it might be traditional to have the degree of anatomical change or the elapse of time since branching determine the length of the branch. This shows his conviction that all that is important are branch points and relationships. We just need to know if a shared modified character exists, not how much it is modified.

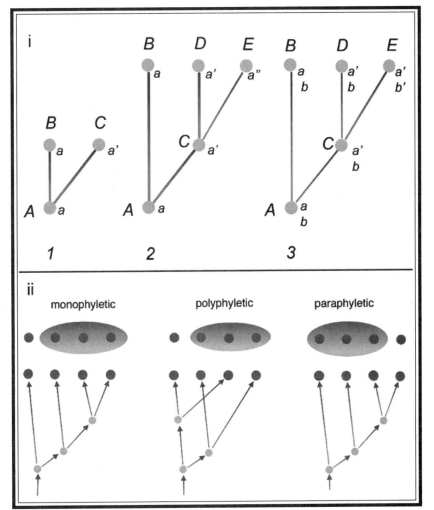

Figure 5-3(i): Hennig's methods for cladistic description of lineages. A cladogram is a branching tree that represents the relationships among species. When character state a' occurs in species D and E, where their common ancestor had character state a for that character, then we can call D and E sister species, and call A the stem species. In this system we call a the original or plesiomorphic state and we call a' the derived or apomorphic state. (ii) Hennig is also concerned with the meaning of higher group names such as genus, family, order, and so on. In his preferred monophyletic groups there is a common ancestor, and all descendants are included.

Avoidance of General Morphology and Key Features

Hennig was focused on rejecting two other widely used approaches to classification that are slightly different from his approach. In one method, the scientist tries to assess the overall morphological similarity between two species. The ones that are most similar are judged to be the most closely related. Overall similarity can be determined subjectively, or by measuring multiple features and using a computer method (phenetics) to produce a quantitative measure of similarity. Either way, Hennig feels that this will produce wrong answers, as it was known that the rate of change varies greatly among species. Even if species E in the example above had been dramatically altered in developing a", it would still be the sister group to species D, no matter how similar species D was to species B in overall appearance. The other approach he was eager to reject relied on the appearance of key features. For instance, when feathers first appear, Hennig is in no rush to declare the emergence of birds. If a feather is found in any dinosaur that is overall still a dinosaur, then we might be able to use feathers to show the relationship of that dinosaur species to its sister species, but that is the end of it. In this case, we have a *symplesiomorphy* of no value at all in determining relationships. We cannot achieve any utility in classification by merely stating that "birds are a large array of species that have feathers."

There are two other reasons why Hennig wished to discourage the use of key features in classification. One is that they may emerge more than once by parallel evolution. If something is extremely beneficial, it is common for related groups to evolve the same feature independently out of a shared predecessor character. The other is his conviction that we will produce the most accurate cladograms if we use as many different sources of data and characters as possible to accurately determine the branching point. He felt it was wrong to pick out a single key feature when other less striking features might add up to a different result.

Monophyly and Paraphyly—Doing Away With Invertebrates, Reptiles, Dinosaurs, Monkeys, and Apes

So far, we have covered two of the three major areas where Hennig's view has upturned so much of zoology. He has no interest in the biological species concept championed by Mayr, and no use for the key features that have been used to define the various types of creatures. Moreover, his system requires the abandonment of the concepts and names for a variety of animal groups, including invertebrates, reptiles, ungulates, monkeys, and apes, among others. Not only are the names themselves problematic, but Hennig argues that these groups do not exist because they make no sense within the cladistic frame of logic. There are alternative terms for groups that cover some of this ground, but they are terms that tend to be unfamiliar to the general layperson, and they don't really carry the same biological meaning. The classic example concerns reptiles, mammals, and birds. Hennig suggests that we must only have monophyletic groups (those coming from a single ancestor), and that you can only refer to a monophyletic group that includes *all* of its descendants (see Figure 5-3ii). That sounds reasonable enough, but look what happens.

New Names and New Groups to Fit the Principles

Out of the ancient amphibian group, the common ancestor of the reptiles and mammals emerged. The synapsids split off to eventually lead to the mammals, the anapsids lead to the turtles, and the diapsids split off to lead to the remaining reptiles. In the reptile group, later splits led to snakes, lizards, and archosaurs (crocodiles and dinosaurs). Then, out of the dinosaurs, the birds emerged. Now, we commonly speak of mammals, birds, and reptiles. By reptiles, we mean turtles, lizards, snakes, crocodiles, and dinosaurs, but we certainly don't mean to imply that birds are reptiles. Typically, reptiles have scales and are cold-blooded, whereas birds are warm-blooded, are feathered, and fly. It seems safe, sane, and appropriate to rank birds as a separate group on a par with mammals and reptiles, but that is a violation of Hennig's principle of monophyly.

From Hennig's point of view, the reptiles group is not monophyletic; it is called paraphyletic because we exclude two major groups—the birds and the mammals—from inside the reptile clade. We can refer to diapsid reptiles and birds as the *Diapsida*, but this is my point about unfamiliar terms that mean something different biologically. We can say turtles, lizards, snakes, crocodiles, and dinosaurs every time we want to say reptiles. Or, we can say "non-bird, non-mammal reptiles." But we can't say reptiles. In fact, because mammals came out of the synapsid reptiles, it is best to use the term *amniotes*, which includes all conventional existing and extinct reptiles as well as the mammals and birds.

In a book such as this, one meant to be accessible to the general reader, this is a problem. When a biologist uses non-Hennigian terms, many zoologists will infer that the scientist is just too simpleminded to understand Hennig's logic. The problem with the term invertebrates is that it is paraphyletic, because it omits the vertebrates that come from within the invertebrate groups; therefore it's much better to say bilateria. We can't say dinosaur because that omits birds; better to say archosauria. We can't say ungulates or artiodactyls, as these terms omit the whales and dolphins; cetartiodactyla has been suggested, although that clearly is a word that refers to a clade or lineage rather than to a type of animal. Finally, apes and humans split off from the monkey group so we can't talk about monkeys; instead we should say haplorrhines or simiiformes. In fact, because humans split off from apes, we can't use the term apes at all—and that goes right to the title of this book! Apes are paraphyletic because it is a group that omits humans. Instead, we should say hominoids.

The Debate Over PhyloCode

However, before we go too far with revisiting the new names, we have to accommodate the looming onset of the PhyloCode. Because all we have left in the end is a vast array of cladograms, and because the only features of concern are those that determine sister species branching, there appears to be no justification and indeed no need for any names at all, other than species names. We can keep traditional names such as *Homo sapiens* by converting them to one word (Homosapiens) or by abandoning the Latin and the genus name (species: thinking). Better yet, a simple bar code or numerical system as there is for galaxies might be more Hennigian.

Hennig objects to paraphyletic groups on a number of grounds (see Hennig 1966, 238). Some of his objections are a bit facetious. For instance, he suggests that if we allow a professor to talk about a group called the birds that is fine, but because it is that individual's subjective judgment that feathers, warm-bloodedness, and flying make up something special, someone else might not agree and confusion will result. He also points out that it alters the meaning of extinction; thus, we don't have a basis to say dinosaurs are extinct. Archosaurs never went extinct, but we now have many bird species, but none of the ornithischian archosaurs and none of the non-bird saurischian archosaurs.

Hennig and Cladistics Remain Controversial

Ernst Mayr was very upset about all of this name shifting:

Wherever monophyly is discussed, the treatment is marred by the self-serving claims of the Hennigians. The simple fact is that for 90 years the term monophyletic has always referred to the status of a taxon. It was Hennig who shifted the term to an entirely new concept. Such a transfer of terms is simply not permissible according to all the traditions of science. There is no reason whatsoever not to accept a term for Hennig's new concept, but that term is not monophyly, but holophyly (Wheeler and Meier 2000, 96).

It's not just the usage of the term *monophyly* that Mayr had a problem with. He goes on to attack the entire cladistic program:

> The terms apomorphic and plesiomorphic are excellent choices for describing characters of taxa in a phyletic lineage. However, they are, in my opinion, non-sensical when applied to species (Ibid.).

Ernst Mayr seemed to be mystified as to why so many biologists pay so much attention to Hennig. Certainly, part of the attraction is the appearance of logical clarity and simple newness. One reason why Hennig's use of the term *monophyly* has been so effective is exactly what Mayr is pointing out: In the old days before Hennig, a biologist trying to complete a classification of a group would be widely considered incompetent or heretical if his analysis produced a group that was later proven not to be monophyletic. However, in those days, the opposite of monophyly was polyphyly (multiple different ancestors contributing to the group). Carl Linnaeus did the first major correction of a polyphyly error in the 1770s when he moved whales from the fish and into the mammals. Hennig's "paraphyly" sounds like "polyphyly," but it really is a whole different matter.

Hennig's concept concerns not only the origin of the group from a single ancestor, but also the fact that we cannot remove a group—vertebrates, birds, cetaceans, apes, or humans—from the ancestral clade for useful biological purposes just because a random biologist thinks those groups represent something special. Nonetheless, when Hennig instructs us that it is misleading to consider a group called the dinosaurs and to say they are extinct, I agree with Mayr that Hennig's approach can be more of an impediment than an aid to understanding the evolution of organisms.

Mayr accepts that *holophyly* (providing a name that includes all the descendants) may be helpful for some areas of systematics. However, he feels that language is most effective when it is adapted to the realities of biology. Language needs to be available that readily signifies that the origin of groups such as the vertebrates, birds, cetaceans, apes, or humans represents significant transformations. This is clearly a critical property of evolution. We don't have anything as simple as a vast array of slightly different species of blue-green algae making up all of life. Rather, we can see that evolution periodically produces a group that is quite different from its ancestors. The ancestral group continues to be what it was—dinosaur, artiodactyl—but the new group—bird, cetacean—represents a difference in body plan or adaptive framework that is a true innovation in nature. Cladistics may not have or even need any language for reflecting this property of evolution, but biology in general does demand such terms.

Treating Morphology as Mysticism

A key element of the Hennigian program is the deemphasis of morphology. This position is on a collision course with embryological genetics. Feathers or no feathers, the axial High Order Modules (HOMs) in the ancestral bird underwent a dramatic and sudden modification. This resulted in a near doubling of neck length and a distortion of the upper limbs into elongated, flat structures. This critical homeotic transformation is the basis of birds as we know them today. Major changes of the body plan due to significant alteration of the *Hox* array of genes are biologically definite phenomena that create new types of animals. This has happened from time to time in the history of the animals, and has resulted in the establishment of new phyla and classes of animals.

If you cannot tell the difference between a bird and a dinosaur, then you cannot tell the difference between a single-celled, blue-green algae and a human. To say that anything other than counting the number of clade branches is a subjective way to distinguish species is a complete repudiation not only of common sense, but of biology in particular and science in general. Are galaxies different from planets? Is the universe different from a spec of dust? In biology, just as in astronomy, hierarchy is not just a leftover of medieval traditions, but a real property of the biotic world that is useful in understanding evolutionary processes.

As the advocates of PhyloCode head in one direction—abandoning the concept of higher level groups—Stephen Gould leads in the opposite direction. He believed that groups above the level of the species had their own types of evolutionary processes, and that they must therefore be granted support as a matter of biological reality, rather than simply as a matter of historical naming conventions (Gould 2002). One reason why Gould's hierarchical selection and Mayr's biological species concept are ignored by strict cladists is that in may ways, cladistics is not about evolution. It is possible to carry out a cladistics analysis of relationships without any reference to what characters are changing or why. It also really does not require speciation or any other usual aspect of biology or evolution. However, as many involved in systematic biology (the field of classification) are aware, by abandoning the rest of biology in pursuit of logical purity, they risk undermining the entire reason for their own existence.

Hennig and Higher Order Classification

The first distinct member of the vertebrate lineage therefore may well have had the upside-down, inverted body plan that differentiates our phylum from the body plans of the other bilaterian phyla. When that happened, would that individual and its new progeny be just a new species, or would it be a new genus, a new family, a new order, a new class, and a new phylum all at once from its first day? From Ernst Mayr's definition of the biological species concept, we can certainly say that this strange new creature and its immediate descendants did represent a new species. However, classification terms above the level of the species have not seemed to have the underlying real basis in biology as the species.

Body Plans and Higher Level Classification

An important clue here can be seen in the concept of the body plan. This has been based on a subjective impression of the difference in design that can be seen between major different groups of animals. When you compare a sponge, an octopus, a fruit fly, a worm, and a lion, it doesn't take much insight to see what is meant when we say that each has a different body plan. But is there a strict biological basis for assessing this?

One option is to look at differences in the *Hox* gene array (see Figure 7-3). We don't have all the information yet, but it seems that in many cases, the significant changes in body plan have been accompanied by significant changes in the detailed organization of the *Hox* genes. As long as we adhere to strict Darwinian and Modern Synthesis views of how change takes place in biology, we don't have to ask the question about definition of a phylum, class, order, family, or genus. If change is gradual and steady, then we expect the original species of the vertebrate lineage to be very similar in appearance to the parent species from which it separated. The difference in appearance between a modern mammal and an insect would just reflect the very great time interval—500 million years—since their separation and the steady accumulation of differences. However, if dramatic changes in body plan happened almost immediately at the start of the vertebrate lineage, don't we need terms (such as phylum or class) to describe this very large change as soon as it first occurred?

A Proposal for Biological Definitions of Higher Level Groups

If a new phylum can appear as rapidly as a new species, it is helpful to have some biological definitions. Even without the full details of the *Hox* genes from a large number of animal types, and without the details of the *Hox* arrays present in the ancient lineage ancestors of 500 million years ago, it is possible to look at changes in the bilateria that affect their axial High Order Modules (HOMs, or segments along the longitudinal axis) and deploy them to demarcate categories of classification at least at the level of phylum (Chordata, Arthropoda) and class (fish, amphibian, mammal). In many cases, differences in the appendicular HOMs (limbs) are used to define and categorize animals at the level of the order—for example, Carnivora for lions, cloven-hoofed artiodactyls for antelopes, single-toed perissodactyls for horses, and Cetacea for whales. Within the orders, major changes in the skull seem to be used to define the family (canid, felid, hyaenid). Assignment of the genus typically involves body size and dental specialization (lion vs. cat). This is not a concrete breakdown, but, as we learn more about morphological genetics, this sort of scale may help assign specific categories of genetic change to specific levels of hierarchical classification, at least as it applies to vertebrates.

Another approach uses the methodology of a field called *phenetics*, a mathematical study of animal form. In this method, we measure a great number of features of a several animals, throw all the data into a computer, and come up with a numerical summary of how much change in shape has taken place between and among the groups measured. If the amount of genetic change is small, but the amount of total phenetic morphologic change is large, then we can suspect a highly pleiotropic change (when multiple features are affected by a single gene) has occurred. A ratio between the amount of phenetic change as the numerator and the amount of phenetic change as the denominator can capture this. A very large number points to a major reorganization, and a hunt for a small but critical change in morphogenetic genes may prove very valuable.

Pleiomorphic changes (those affecting multiple parts of an animal) can come also come from what we call *quantitative trait loci*. These are changes that affect multiple aspects of anatomy by, for instance, making a new species of animal much larger or smaller than its ancestral species—for example, the cat and the lion. In general, there are predictable rules of scaling called *allometry* that show how parts change relative to each other when overall size is different; for example, small animals such as a mouse will have tiny limbs in relation to their overall body size when compared to those of a horse or elephant. When effects of this type of change are cancelled out in the phenetic analysis, the results can show the amount of actual reorganization or morphologic innovation that the genetic change has accomplished.

From the point of view of the cladists, this analysis is completely unnecessary. Willi Hennig pointed out that we should simply plan to decipher an endlessly branching tree; all that is important is that we get the relationships among lineages correct—that is, the process and result must be truly phylogenetic, or showing a true reflection of genealogical lineages. However, to describe and organize the biotic world and to better understand the way in which genetic change interacts with selection, we need analytic tools and classifying language that aids communication about this central aspect of biology.

Summary

Played out to its extreme conclusion, a completely pure cladistics analysis with PhyloCode naming will eventually produce a vast uniform branching tree showing a location for every one of the millions of species. Because each node on the tree will have a number, we will be able to abandon traditional animal names. Furthermore, we will not need to make any special notation relating node number to the shape or appearance of the organisms that are at each node. The tree and ID code will not provide any distinction between, say, bacteria and Bactrian camels; each will have a location on the tree and a similar numerical code. Fortunately, the rationale for the oddities of this formal cladistics enterprise have been undermined by morphologic genetics. Not only does the shape of an organism have an objective basis in genetic code, but we have an objective system for ranking the relative importance of various types of change. A reordering of the *Hox* genes and body segments merits the naming of a new high order category.

This new analysis helps provide the objective basis for what biologists have done for many years—namely, the study and description of different types of organisms, and how their shapes and biological functions interact with their environments. The interaction of environment with genes is structured in this way, and this is truly the basis of evolutionary change.

Modular Selection Theory: Evolution at Multiple Levels

Introduction

A critical element of Darwin's theory of evolution is the idea that within a species, too many organisms will be born for the environment to support them. Given a background of variation among the members of the species, some—the ones that are better adapted to their environment—will do better than others in the struggle to survive and reproduce. However, there is growing evidence that competition may take place at higher levels—for instance, the mammals vs. the dinosaurs. In addition, competitive selection appears to take place at levels below the species. Most important for animals, the High Order Modules (HOMs), such as insect segments and the somites out of which vertebrae emerge, may effectively compete with each other to generate new body forms (think of the lizard and the snake).

This Modular Selection Theory, as it is called, provides a major extension of the Darwinian concepts of evolution. It is important because it brings a wide array of biological phenomena into the evolutionary framework that otherwise cannot be accommodated. Ranging from competition among phyla in oceans, to competition among enzyme clusters in the cell, the application of evolutionary models of adaptation, selection, and change is revolutionizing our understanding and scientific framework throughout biology. First elaborated in detail by Stephen Jay Gould in his final 2002 magnum opus, *The Structure of Evolutionary Theory*, the full details and implications are just now beginning to be explored in detail.

Individuality Theory
Ghiselin and Hobbes—A Broader Definition of the Individual

Gould's concept of selection at multiple levels depends in significant part on a new understanding of what the word *individual* really means. Michael Ghiselin, a philosopher of biology, first raised this seemingly simple issue 30 years ago (Ghiselin 1969). An historical aside about the redefinition of this word in the field of politics may be helpful in gaining a firmer understanding to carry forward through the rest of the chapter.

After nearly a century of political calm following on the celebrated rule of Elizabeth I (who reigned from 1558 to 1603), Britain descended into a series of internal battles, including three civil wars and two more wars among the kingdoms of England, Ireland, and Scotland—all in the space of about 12 years between 1639 and 1651. Bishops and noblemen were executed, members of parliament led military factions, entire juries were tortured by the Star Chamber, a king was put to death for treason, and the men who killed the king were then executed.

Fleeing the chaos, and fearing that his own association would make him the next to fall to the executioner, Thomas Hobbes traveled to the relative calm of Paris to continue his great project of designing and explaining a logical basis for governance and politics. The most famous product of that effort is a work that is widely accepted as the basis for all modern writing about political philosophy in the West, and that served as the touchstone for later works that underpinned the emergence of new forms of government in Europe and North America. In that work, *Leviathan*, Hobbes suggests that for man without organization or government, life will be "nasty, brutish, and short." The principle sentiment of his work is captured by the remarkable lithograph that serves as the frontispiece for the book (see Figure 6-1). The illustration likens the concept of the state to a giant individual made up of its many subjects.

On the political side, Hobbes argued that men should willingly make a contract in which they surrender all rights to commit violence against each other in favor of an absolute monopoly on the use of violence to be held by the king. The citizens would retain control of the consent to be governed and would therefore be able to create or terminate the government based on their approval. Additional populations could be added or divided out—for example, Scotland or Ireland—based on the consent of those citizens.

Individuality Theory and the Species

With no stated awareness of or attention to Hobbes or his drawing, Michael Ghiselin launched a philosophical revolution in evolutionary biology with his 1969 book (written when he was freshly out of graduate school), which pointed out that it would be correct to think of a species as if it were an individual itself. A species would be born (speciation event) and it would someday die (extinction). It could reproduce (produce daughter species), and it would only exist as long as it maintained sufficient geographical continuity to maintain gene flow among the organisms out of which it was constituted.

From a philosophical or metaphysical point of view, Ghiselin has been concerned with erasing the remnants of platonic idealism from the process of defining a species (Ghiselin 1997). For Aristotle or Linnaeus, it was appropriate to state a number of features or characteristics that captured the essence of what was unique about the species or even an individual, representative type specimen. For example, a typical adult male human is a tailless primate with minimal body hair, habitual bipedal locomotion, opposable thumbs, and a heavier brain (usually about 1350 ccs) than those of apes, averages 5'10", 160 pounds, and uses complex language. With this definition in hand, a zoologist could identify similar creatures as belonging to the species *Homo sapiens* whenever they were encountered. If they differed from the description, the zoologist could check the definition of similar species and decide if the creature was close enough to fit into the same species or if a new species needed to be named. This is *typology*, the description of an ideal version of a class of similar objects or organisms. Sir Richard Owen's archetype (see Figure 1-2 in Chapter 1) was an example of an ideal type specimen for the vertebrate animals.

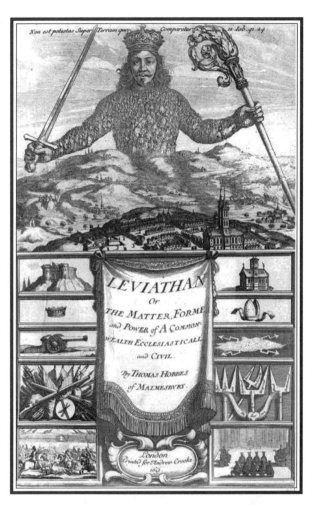

Figure 6-1: Hobbes's *Leviathan*. The concept of a species as an individual made out of many organisms that are species members is captured in this frontispiece from Thomas Hobbes's famous book on political philosophy, *Leviathan* (1651).

Biology indicates, however, that the organisms in a species are all a little different from each other, with the exception of some identical twins. There does not seem to be any valid biological basis for picking out one perfectly representative single organism and calling it the perfect representative. All the varying members of the species are equally entitled to be seen as representatives of that species—tall, short, red hair, blonde hair, dark skin, light skin, thin, heavy, eloquent, mute, and so on. Basically, philosophy reveals that you can describe a species, but you cannot define it. The same area of philosophy reveals that the opposite of a class (defined by a type specimen) is an individual (a single entity with a variety of describable features). It is most helpful here to keep Thomas Hobbes's illustration in mind. Consider the species as a kind of individual made up of the numerous organisms that are members of the species, just as the frontispiece drawing shows a nation state made up of its citizens.

Individuality of Higher Groups

Ghiselin's primary concern was philosophical in that he wanted to help move evolutionary biology away from 19th-century typology, and toward a more modern and scientific empirical footing based on observation and description. His proposal has not been completely accepted among biologists, in part because it does represent a significant change of thinking. However, Stephen J. Gould pounced on the idea to suggest a widespread reinterpretation of the way in which evolution does its work. Darwin was concerned with competition among the varying members of a species. Some had better fitness and would be more successful at surviving and reproducing. Natural selection therefore acted at the level of the organism to gradually change the genetic makeup of the entire species population.

Gould followed Ghiselin's philosophical clarification by pointing out that, if species were viewed as individuals (think again of the Hobbes drawing), there might be competition among closely related sister species. Rather than limiting the effects of natural selection to determining the reproductive success of organisms within a species, couldn't natural selection act at the level of one species competing against another? This led Gould over a number of years to finally propose a hierarchical plan for natural selection and evolution. This was fully articulated only in his final, massive 2002 masterpiece, *The Structure of Evolutionary Theory*, published just six weeks before his death at age 60. The great length of the book suggests that he had all the time he needed to complete it. The ideas reflect a summary of his career of nearly 40 years describing many of the intellectual wrong turns and mid-course corrections that led him to his final set of theories. However, from following his work and studying with him, it is clear to me that the some of his newest ideas could have benefited from further revision if he had had more time.

Selection at Multiple Levels

At the core of what Gould had to say are his ideas about *macroevolution*, or evolution *above* the species level. Why stop with identifying species as biological individuals? At a higher level, genera, families, orders, classes, and even kingdoms might usefully be viewed as individuals having a birth and a death, variation from sister groups, and relative success at survival based on their fitness relative to other higher groups. He points out that this concept works at levels of organization below the level of the organism as well. In particular, he points out that even cells are capable of being viewed as individuals, as most of the cells in the world actually are unicellular animals, plants, and fungi (paramecium, algae, plasmodium). Below the level of the cell, he accepts that genes may be viewed as individuals subject to selection, but he differs sharply from Richard Dawkins's famous work, *The Selfish Gene*, in regard to the details.

Dawkins had achieved considerable success with the claim that our bodies are mere vehicles, and that it is the genes themselves that are the focus of evolution. Gould accepts only a very limited role for evolution of genes as individuals. We know that within the genome, some genes can become duplicated. This can happen again and again, and many important genes clearly come to be represented in multiple copies within the genome. The various copies can become different from one another. If one version takes over as the more effective one, the other versions may eventually cease to be transcribed into proteins. As they become vestigial and unimportant, evolution subjects them to random change until they are effectively dead genes: mutilated, virtually unrecognizable, and incapable of directing the assembly of a working protein. This is very different from Dawkins's view that everything we see upon the great stage of nature is a matter of the evolution and selection at the level of the gene. In fact, most of the competition among genes will go forward without any detectable effect on the organisms in which they reside.

Success in Species Level Competition

When Gould considered evolution and selection as they pertain to cells in an organism, I think he missed the point entirely by setting the wrong benchmarks for success. When an organism that is a member of a species succeeds under natural selection, it survives to reproduce and produces more offspring than its competitors. When a species is successful against other sister species, it might generate more daughter species, but this seems to happen at a very slow rate. Look at the example of *Homo sapiens* and its success over a few hundred years. What has resulted is a phenomenal increase in the number of organisms making up our species. There are only a few hundred gorillas living, all in a few small areas of the world and relatively near each other. However, there are billions of humans spread through nearly every environment on the planet.

An individual organism can only grow a little bit as a token of it success (for example, growing to be full size and maybe putting on a little weight). An individual species, on the other hand, can grow in size by many orders of magnitude. There are nearly 10 million humans alive for every gorilla in existence. Any of the higher groups can grow in this way. A vast array of dinosaurs replaced the earlier land tetrapods, and vast array of mammals replaced the dinosaurs—a process in which extinction overtook every single species bar none.

Success in Selection Below the Level of the Organism

At the level of genes, neither vast increases in different daughter gene versions in the genome nor vast increases in copies of any gene take place. However, success for genes may be measured by assessing the effectiveness of one version of a gene relative to the sister version. Here the effectiveness can be assessed in the contribution of the gene to the fitness of the organism in which it exists. The comparable situation for cells in an animal is more starkly apparent and highlights Gould's error. He states that within a multi-cellular organism, evolution of cells must be suppressed. The concept is that one type of cell will grow in prevalence to take over the body from all the other types of cells. He points out that this is called cancer, and that it is self-defeating evolution because in its success there is a pyrrhic victory: The organism dies, killing all the cells along with it.

Gould's personal history actually plays a role in this view. He came to Harvard in 1967 having already gained considerable attention with his early work in rates and patterns of change among fossil species. He was rapidly promoted, as the lightning rod effect of his first major theory in macroevolution, punctuated equilibrium, made him a central figure in evolutionary biology. Aside from his series of 300 monthly articles for *Scientific American*, his appearance on the cover of *Newsweek* magazine—in a piece not unlike the famous *Time* cover announcing Bruce Springsteen as the new Bob Dylan—helped seal his position as the best-known evolutionist of his time. Among his greatest pleasures was the opportunity to move into the grand office in the Museum of Comparative Zoology that had once belonged to Alexander Agassiz, whose father, Louis Agassiz, was the founder of not only the Museum but of American paleontology in the 1850s.

Surrounded by his huge collection of books and specimens, he produced a seemingly endless series of books and articles redefining his field. Unfortunately, the specimen cabinets in the nearby basement were full of asbestos board. In 1982, he contracted an asbestos-related cancer, mesothelioma. Although this is an almost universally lethal cancer that kills in months, Gould was cured and continued writing with undiminished vigor until he contracted a lung cancer that was most likely asbestos-related as well. It was this disease that killed him in May of 2002.

Forgiving Gould his focus on suppression of excess cellular growth, an alternate view of success for a lineage of cells in a multi-cellular organism would be an increase in functional efficiency that helps the cell type or tissue become more important to the organism. A good example is the diversification of blood-cell types. In insects and other invertebrates, there is relatively little role for blood cells; in vertebrates, however, there are several types of blood cells. In most mammals, there are many different subtypes of white cells with various immune functions as well as highly efficient oxygen carrying red cells. It is conceivable for a new group of invertebrates to arise without blood cells, but it can be reasonably assured that this will not happen among the mammals. Similarly, some of the gut lining cells of fish become specialized (adapted) for gas exchange as they come to line the lungs of various land animals. Mesodermal cells begin to produce subtypes of cartilage and bone cells to provide an endoskeleton, and skin cells become impermeable so the creature can move away from dry land. If we restrict evolutionary success to competition among cells within the body, the result must still be diversification of cell lineages in a way that is analogous to the successful production of numerous daughter species, even if the total number of organisms in an environment remains constant.

Effects of Selection in High Order Modules (HOMs)

An interesting thing happens when we look to the highest level of internal organization within an organism, the level that Goethe and Geoffroy were focused upon. The ultimate example is the *metamer* in the body, such as the vertebra or insect segment. The somatic segments of vertebrate animals begin to form a short time after the organism commences its embryonic development. Through the process of terminal addition, multiple copies of the basic segment are formed. The number of segments formed can be greatly increased, although there appears to be an empirical limit of around 250 segments. We have insects and millipedes, fish and eels, frogs and caecilians, lizards and snakes. In each case, a major group has emerged with greatly increased numbers of segments. At another level, we see specialization of segments and effective competition. Tail segments become less important, and are often reduced and abandoned, and thoracic segments, which are important for respiration, become progressively specialized. The thoracic and lumbar segments together play an important role in locomotion.

The Structure of Evolution Below the Level of the Organism

In order to properly consider the role of evolution at levels below the organism, we have to accommodate the relationship between our chosen level of subdivision and the genome. At levels above the organism, such as the species, genus, or family, selection may act upon the entire group (clade), but the repository for the genetic information that will generate, store, and propagate the advantageous changes still lies in the genome of the organisms that make up the clade. Similarly, for evolution of multienzyme conglomerates, cells, tissues, organs, or High Order Modules, the relevant information is still stored in the genome of the organism.

When examining evolution at or above the level of an organism, it is readily apparent that the genome for each organism will be capable of directing the assembly of phenotypes and morphologies in the organism in which it resides, and in the clades in which we consider it. However, when working below the level of the organism, this is not necessarily the case. As Gould points out, what is important from the point of evolution is any genetic change that can be passed along in the germ line that carries the organism's genome forward to the next generation. In cancer, the genome of an individual cell may respond to an environmental stimulus by directing it to multiply aggressively; as with Professor Gould's mesothelioma, this may or may not kill the organism. However, for a variety of reasons, this has no evolutionary effect. The genetic alteration in that cell cannot be passed on to posterity. This points out a fundamental requirement for evolution below the level of the organism: Changes must be heritable and capable of being passed along to descendant organisms, and they must be capable of identifying a specific target agent that either participates in evolution by interaction with the environment, or is capable of competing with other components inside the organism. For instance, there is no conceivable genetic change that will identify a particular individual cell in the wall of the gut—among the millions of nearly identical cells around it—and cause that one cell to act in some unique way. Yes, the copy of the genome inside that one cell may experience a change that causes cancer, but the heritable genome stored in the germ cells does not seem to be capable of identifying individual cells. Rather, the information in our genomes is directed at the somewhat higher level of cell types and tissues.

Following along this line, it is possible to conceive of, for example, a genetic change that caused an improved digestive enzyme to be secreted by every stomach lining cell. Eventually, if the stomach cells achieved astonishing efficiency in causing complete, accurate, and immediate digestion of all foods with perfect, rapid absorption, then there would be no need for gut-lining cells in the small or large intestines (and indeed, no need for the rest of the gut at all). The animal could have a 100-percent-efficient stomach and a connection to an outlet for waste, but rest of the digestive system could be dispensed with. This would be a heritable change, one that identified a subset of cells and allowed them to displace all related intestinal lining cells. There might or might not be any benefit to the entire organism in this, as the same ultimate effect would be achieved: The animal would accomplish digestion of what it had eaten. But why stop with digestion? What if the intestinal lining cells could also accomplish gas exchange? If an animal just swallowed (or breathed) air into its gut then the gut lining cells could take over gas exchange— perhaps in a new air sac structure. In this fashion, these cells could leave the gas exchange cells of the skin and gills in the dust so that the organism would use only intestinal lining cells for digestion and respiration. This is actually how lungs in reptiles and mammals displaced gills and gas exchange through the skin, as it occurs in amphibians.

The first of these examples of cell-type evolution (gut lining competition) has not occurred, but the second example (take over of gas exchange by gut cells) certainly has. In fact, for many tissues in the body it is possible to draw out branching tree diagrams that show the descent of one tissue type from another. This process of progressive subspecialization of cell types of the various tissues can also be observed as it unfolds in the embryo.

Hox Genes and the Selection of High Order Modules (HOMs)

Evolution of cell types seems to make sense from this point of view. However, what are some examples at higher levels within the organism? This is where we come to what may be called High Order Modules, or HOMs. Three important examples of HOMs are the segments of bilaterian animals (insects, crustaceans, vertebrates), the limbs of vertebrates, and the teeth of mammals. The limbs of vertebrates have a genetic control system for their embryonic assembly that echoes the assembly of the main body segments, and there is a similar but less complex system for mammalian teeth. However, the role of evolution of HOMs is best understood by attending to the axial segments (such as those marked by their skeletal component, the vertebrae) arrayed along the length of bilaterian creatures. These axial HOMs are typically composed of a variety of tissue types, including skin, bone (or exoskeleton), muscle, nerve, circulatory, digestive, and respiratory components. This is entirely obvious in insect segments, but is most clear in vertebrates when the embryological progression of the *somites* (embryonic body segments) is considered.

The cell types making up one segment are more or less identical to the cell types making up an adjacent or distant segment along the body. However, unlike the unidentifiable individual cell, each segment can be identified by the morphologic homeobox labeled *Hox* genes that define it (see Figure 7-5). Therefore, these axial HOMs meet the criteria of being subject to individual identification in the genome. Because of this, heritable germ line changes can occur that alter one segment in a particular way relative to other segments around it. This is the special function of the *Hox* genes.

One of Gould's principal criticisms of the so-called selfish gene of Richard Dawkins is that duplication of a gene may not have any effect on the organism (in Dawkins-speak, the "vehicle") that interacts with selection in the environment. For this reason, Gould argues that gene duplications reflect only competition with other bits of DNA in the genome where there may be some limit on tolerable total genome size. The situation with *Hox* genes is quite different. A cursory glance at Figure 7-5 will show that duplication of *Hox* genes can be reflected directly in the external anatomy of the organism. These gene duplications and gene specializations can have a direct interaction with selection in the environment. Not only does the *Hox* gene system allow for increase in number of axial HOM segments, but it also allows for the specialization of different segments for different tasks in the adult organism (as mapped out by differentiation and specialization of the *Hox* gene relevant to a given body segment).

Similar to the example of the cell types, it is also possible for different body segments to become competitive. In fish, all of the body segments more or less do the same thing: They carry muscle attachments for swimming, and they encase the neurologic and digestive tissues. In land vertebrates, the anatomy of some segments (dorsal or thoracic segments) function in respiration; other segments (the tail or caudal segments) remain only marginally critical to the organism, as they are engaged only in moving the tail for balance. In some species, the tail segments become more critical; this includes the spiked tail of the stegosaurus as the animal's most effective weapon, or the prehensile tail of the spider monkey that becomes fairly critical in locomotion and feeding. Nonetheless, time and again, the tail segments are summarily disposed of without much apparent functional detriment to the species—in apes and humans, for instance.

Evidence of Evolutionary Events Among HOMs

Axial HOMs can increase or decrease in number, and they can become more or less specialized relative to each other. They can also be modified in shape to affect other generalized functions of the body. One of the fascinating aspects of the genetic control of axial HOMs is that, despite specialization of body segments in mammals, there is still clear evidence of genetic changes that affect all the body segments at once. This is an example of a pleiotropic change (one genetic alteration affecting many locations in the organism) that magnifies or amplifies the effects of evolutionary changes in the *Hox* genes. To reshape the chest and alter breathing, for example, it is not necessary for each of the body segments to independently evolve a similar helpful change. Rather, all of the ribs and vertebral extensions can be reshaped in concert by a genetic change that acts on all the segments.

In fish, the ancestral shoulder girdle is attached directly to the skull so that the head cannot be moved relative to the rest of the creature. In amphibians, the body has some separation from the skull, and there is a single cervical (neck) vertebra that allows for some head movement (think of a frog looking up or down). In mammals, lizards, and dinosaurs, there are generally six to 10 cervical vertebra, which enables them to move their heads side to side and up and down for defense, attack, feeding, or general inspection of the surroundings—all without altering their stance and, more importantly, without altering their walking or running gait. Here, increase of the number of specialized cervical axial HOMs has a specific signature in the *Hox* genes, and has a clear and powerful functional benefit that is only abandoned in such rare cases as the whales and dolphins, wherein the neck has not been completely maintained.

Is there really an effect on fitness of the organism if there are 178 instead of 177 segments (as there are in some snakes)? Furthermore, can we really speak of evolution at the level of these axial HOMs? Isn't every anatomical feature of the body subject to evolutionary change? Must we therefore consider a separate level of evolution for mandible shape, claws, hair color, and so on? The unique features of axial HOMs that make them subject to acceptance as a significant, distinct level of selection is that within an individual, they are multiple, they are individually addressable by distinct genes, and they have specialization relative to each other. Because the segments are specified by particular genes, they also have traceable branching genealogies that can show what segment is ancestral to a given pair of existing segment types.

The Explanatory Power of Modular Selection Theory

What is the value of considering all of the various hierarchical levels of evolutionary selection that Stephen Gould has proposed? One of the most important consequences is that it reveals explanations for phenomena that cannot be readily explained when only the single Darwinian level of selection among organisms in a species is considered. At a higher level, we may see why the mammals replaced the dinosaurs, even though a given species of ground sloth, for example, may have been no match for a dinosaur. As we will see in a later chapter, mammals run much faster than dinosaurs, and can access far greater amounts of energy as they do so. Because of this, there was virtually no dinosaur that could catch a mammal fleeing from it, and conversely no dinosaur that could escape a carnivorous mammal by running from it.

Partly because there has been no real theory for clade level selection (for example, mammals vs. dinosaurs), it is widely assumed that it is unworthy of investigation. Instead, a mass extinction due to a meteorite impact that somehow kills off all dinosaurs while sparing most mammals—despite overlapping size and physiology—has been accepted as the basis for biotic change. Although this event may have played a role, a theory that allows for selection at the level of the higher clade opens the door for applying evolutionary mechanisms to this astonishing biotic change. There is no definitive evidence that all of the dinosaurs were extinct before the mammalian radiation of new species began. In fact, it is more out of habit that the possibility of competitive success by mammals has not been fully addressed (see Figure 6-2 and Penny and Phillips 2004).

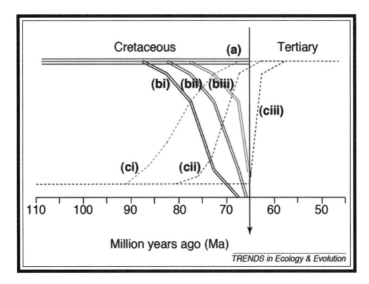

Figure 6-2: The dinosaur-mammal transition—alternate scenarios.
A) Catastrophic: abrupt extinction of all dinosaurs 65 million years ago (a) from meteorite strike, followed by the rise of mammals (ciii).
B) Non-competitive: dinosaurs declined gradually over 20 million years (bi, bii, biii), opening the niches that mammals then filled (ciii).
C) Competitive: gradual decline of dinosaurs (bi, bii, biii) takes place simultaneously with the rise of the mammals (cii) during a 20-million-year period of competition.
D) Competitive 2: major diversification of the mammals (ci) before the dinosaurs decline.

In addition to the need for someone such as Gould to organize and propose a hierarchical theory of selection, there has also been another persisting concern regarding proposals for macroevolutionary change. We can see how a species can change its spots (so to speak). Species member organisms have variation, and some with better spots are more successful at generating offspring. In this fashion, the genetic makeup of the total population of member organisms gradually changes, and the species gains an altered appearance. By definition, there is gene flow among members of the species. However, also by definition, a heritable genetic innovation in one species of a clade cannot be introduced into any other sister species that has already split off. How can the entire clade evolve and adapt to improve its success?

For clade vs. clade competition and replacement, we have two options. Firstly and most commonly, the answer seems to be in the advantage yielded by a shared ancestral feature (*symplesiomorphy*). When the clade first arose and had its initial split into two species, by definition something new had appeared. What we learn from study of the evolution of axial HOMs is that a single gene change can yield a far-reaching and extensive (*pleiotropic*) transformation of much of the body. In some cases (as it was for the then-emerging mammalian lineage), a single gene change can utterly transform the functional efficiency of the descendant clade. In all of the member organisms in all of the subsequent species of that clade, the major advance is likely to still be present. Any species that loses such critical advance is likely to fall victim to competition from its sister species. The entire emerging clade will have improved competitive fitness overall, relative to a more distantly related clade that split off before the important new feature arose.

Simultaneous Change in Multiple Clade Members

Another option that has never been fully considered that can cause an array of changes to occur is an acceleration in the effective rate of evolutionary innovation across an entire clade. This is important when we try to understand the Cambrian explosion. The bilaterians seem to replace the previous ediacran fauna almost by sheer mass innovation. Dozens of new body plans, tissue layers, modes of locomotion, and modes of attack and defense all appear very rapidly. The capability of the bilaterian clade to generate a wide array of new body plans could well have overwhelmed the ability of their ediacran predecessors to respond by evolving as well. However, this cannot be a matter of a series of changes happening in a lineage with new features passed along. Rather, the process of continuous generation of new body plans had to commence at some point and then continue during the Cambrian explosion (anywhere from 10,000 to 10 million years), with new plans continuing to emerge more or less simultaneously in multiple different lineages. Finally, the entire bilaterian clade seems to have settled down with its nice, wide array of body plans, and seen only relatively slow further change for the next few hundred million years.

One major problem with this is that no one has ever proposed a believable mechanism for an environmentally mediated phenomenon that acts upon the evolutionary rate of a particular clade or taxon above the species level.

Part of the problem is the requirement for a fairly advanced understanding of the mechanism of generation of genetic variation. Not only has this mechanism not been well-understood, but in fact it is almost intentionally neglected, both in Darwinian Theory and in the Modern Synthesis (the unification of Darwinian Evolutionary Theory with population genetics).

Variation in Mutational Rates   The textbooks offer an explanation for variations that has been accepted for many years. Gamma rays cause changes in DNA, and there are enzymes that try to correct any resulting mistakes. Further, mistakes in the germ line DNA often lead to the death of the embryo during development, thus ridding the species' overall gene pool of unhelpful changes. However, every so often a change is not edited out, which produces an advantageous small variant. Natural selection is engaged, and the bearer of the variant produces more offspring than other species members. Thus, the new variant undergoes progressive increase in frequency in the species population.

It has been clear for many years that some portions of the genome appear to be subject to more rapid change than others. This has been very important for molecular evolutionists who have compared DNA sequences from species to species to try to trace out the relationships as well as the time elapsed since the ancestor of two modern species split from each other. It is well-known that some portions of DNA change far too rapidly and too chaotically to make any sense out this process. Analysis of these segments can lead to wildly incorrect overestimates of the time of divergence between two lineages. By contrast, other segments of DNA prove to be astonishingly conservative. Some portions of the *Hox* genes and their homeobox signals are this way. There can be so little change between fruit fly and human that, if only this DNA is used, a wildly improbable underestimate of the time since divergence of the two lineages is produced. This has led the "molecular clock" scientists who work in this field to try to use segments of DNA that seem to produce the correct scale of answer. Even if the same segment is used for many different groups and calibrated from good data from the geological record, there is a problem in that the rates of evolution for a given gene do not appear to be truly stable enough to make any completely reliable clocks. Optimal estimates are therefore generated based on a variety of usually reliable segments; the best fit to all the various selected data is considered the correct answer. This has been a particularly huge problem for dating the divergence of lineages at and before the time of the Cambrian explosion. Calibration is difficult, as the difference in types of organisms is as large as the time scale is vast.

Another important effect that has not been understood in the past can throw an additional monkey wrench in the molecular clock calibration process.

The clue to this came from work on the HIV virus that I conducted with Andrew Lever at St. George's Hospital in London and with Geoffrey Harrison in Professor Lever's lab at Cambridge.

Variability in the Fidelity of Nucleic Acid Copying

Another source of genetic change, in addition to gamma rays, is poor fidelity in copying and reading DNA and RNA. Presumably, with the elapse of billions of years and the completion of quadrillions or quintillions of copies, the organisms of the earth would have come up with enzymes that get the copying work done correctly every time. However, this is not the case. The human immunodeficiency virus (HIV) is an important example, as it is known to rapidly generate new strains and variants that enable it to resist attacks from the cellular defense mechanisms of its human victims, as well as any medications and vaccines that are applied. This virus is locked in a life-and-death struggle against everything that all the best technology from a huge number of determined scientists at numerous maximally equipped laboratories around the world can throw at it. There are a variety of reasons that it has evaded eradication, but rapid evolution is one of them.

Do these viruses receive more gamma rays than other organisms? Not possible. Are they more sensitive to radiation? No. The drive behind their rapid evolution is poor fidelity RNA copying. In replicating the genome to make new viruses, the molecular mechanism of this virus makes an uncommonly large number of mistakes. This leads to a plentiful supply of new variants, and this is the raw stuff out of which successful adaptation is achieved.

Background: The Mechanics of Retrovirus Replication

The HIV virus belongs to a category of organisms that form a special case among all living things. Instead of storing its information in its DNA (the long double helix that stores the genetic information in most other organisms), it has its information stored in RNA. Both RNA and DNA play a very important role in every animal cell; however, RNA tends to be a "worker molecule" that carries information around or holds things in place while the animal cell attends to information from the DNA. Nonetheless, many of the lessons from HIV RNA are highly relevant to evolutionary change in the DNA of plants and animals.

In our research, Andrew Lever and I found that there is an external way to control or vary the fidelity of copying of the HIV genome (Filler and Lever 1997). Our resulting report is fairly unique in that there is no other literature out there that deals with this issue in this way. The RNA in HIV, as in any other organism, is not laid out in a straight line. Some parts resemble the railroad-tie arrangement familiar from DNA—two long strands with cross ties—but other parts are bent into folds and loops. The overall shape of the RNA—the folds, the loops, the U-turns, and the railroad tie straight-aways—are determined by the RNA sequence. The RNA is copied by an enzyme called *reverse transcriptase*, which generates a strand of DNA by more or less duplicating the sequence information in most of the RNA. The DNA is then used to trick the human cell. The human cell uses its own machinery to start following the instructions in the HIV DNA copy. This tricks the human cell into producing huge quantities of new HIV viruses using its own resources. As you can imagine, this is the start of a chain of terrible events for the human victim.

Where are the low-fidelity mistakes made that drives HIV evolutionary adaptation? It is in the work of the reverse transcriptase enzyme. It actually works similar to the engine of a train running along the "tracks" of nucleic acid. It pulls in raw materials (the individual bits of nucleic acid) and connects them together to make a copy of whatever individual type of RNA "letter" is in the tracks. However, the train repeatedly gets derailed and falls off the tracks. Whenever the reverse transcriptase enzyme "engine" falls off the tracks, it releases a short segment of incomplete DNA. By collecting these fragments and running them out on an analytic gel, it can be shown that the enzyme falls off the tracks predictably at bends, loops, and U-turns, and it tends to roll along fairly well on the straight-aways. The HIV enzyme complex is able to stitch together the pieces to make a continuous DNA copy in the end, but it tends to make errors at the ends of its fragments as it falls off the tracks.

The result of all of this is that the HIV reverse transcriptase has a characteristic slightly slow rate of copying and a high error rate (aka low-fidelity copying). The discovery from the research project was that unlike a real train that relies on gravity to stay on the tracks, the HIV reverse transcriptase uses electrostatic charge, going from positive to negative, to stay on. This is typical of a wide variety of nucleic acid polymerase enzymes that do various copying and repair tasks throughout the living world. It generates the electrical attraction by effectively grasping a small number of charged metal cations (pronounced "cat-eye-ons"). An example of a metal cation is magnesium. This is more or less similar to the salt dissolved in the oceans of the world, and very similar to sodium (positive single charge), chloride (negative single charge), and calcium (positive double charge). Magnesium is called *divalent* because it has two unpaired electrons in its ionic state as an atom, so it has a net charge of two positive units. With a slight negative charge on the RNA strands (due to their acidic nature), the reverse transcriptase enzyme can use its divalent magnesium atoms to help it stick to the RNA while it does its copying work.

Using special solutions we designed called *chelation buffers*, we found that, for many of the trivalent elements, the reverse transcriptase underwent *cation substitution*. The science here is somewhat complex, but the results were

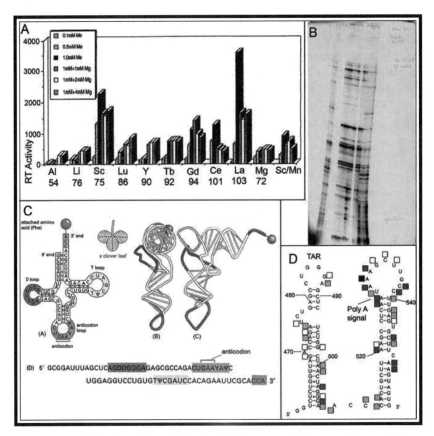

Figure 6-3: Modification of mutability by environmental factors. The rapidly evolving HIV genome demonstrates environmental effects on rate of mutation.
A) The speed of reverse transcription from RNA to DNA is affected by the concentration of various elements in the medium.
B) The polymerase enzyme produces fragments of the genome with standardized breakpoints sorted out by length in electrophoretic gel. Some elements cause improved readthrough—longer segments of DNA so that certain bands disappear in some channels.
C) RNA has a 3-D structure of hairpins and loops. Breakpoints correspond to geometric twists and turns of the RNA.
D) HIV genome shows high rates of transcript breakage at loops and bends.

truly astonishing: The reverse transcriptase was now completing its copies five to 10 times faster than usual (see Figure 6-3A). As we analyzed the effect, we found that instead of making many short pieces that had to be joined together, it was making mostly long, complete transcripts. The engine was no longer falling off the tracks. It was sticking better. We had converted the low-fidelity enzyme to a high-fidelity enzyme. The reason for the increased speed of copying was that the enzyme was now spending all of its time copying, and very little time falling off and trying to get back on again. The task of piecing together the fragments was greatly reduced (see Figure 6-3B).

We had been hoping to find inhibitors of reverse transcriptase for the treatment of AIDS, and we did find some types of cations that did just that. They substituted for the magnesium but apparently provided very little adhesion. The enzyme made only small amounts of short transcripts—basically, it was such a mess that it could not complete its job of copying the RNA. The acceleration discovery has some application in industrial or forensic applications, where it is useful to make copies of nucleic acids, but unfortunately it was worse than useless as a treatment. In fact, we had to do a series of subsequent experiments to prove that no common medications containing trivalents were activating the acceleration effect by releasing the wrong cations.

Polymerase Fidelity, Ecologic Change, and Rates of Evolution

So what does all of this have to do with evolution and the Cambrian explosion? Everything, as it turns out. The experiment showed that an external change in the salt concentration of the ocean can cause a change in the rate of evolution among the creatures that have a polymerase enzyme sensitive to it. Cerium is the most abundant lanthanide element, and is just as common as nickel and copper. Scandium is another trivalent cation that is present at even higher levels than cerium. We can at least propose that various suitable trivalent cations were present in some concentration in the ocean at the time of the Cambrian explosion, and that these are often incorporated in polymerase enzymes at a concentration relative to magnesium, which had been relatively constant for some extended period. The balance between trivalents and magnesium would set the rate and fidelity of many nucleic acid polymerase activities. In the case of a major meteorite impact or a major climate change that temporarily altered the cation contents of the ocean (Stanley et al. 2002), we at least have a mechanism that could lower the fidelity of DNA copying across a group or clade of organisms. This in turn would cause an increase in the production of variants, and would seem to cause evolutionary change to accelerate.

Do bilaterian animals have important segments of loop, curve, and U-turn nucleic acid information that would be sensitive to this effect and thus be important for morphologic evolution? The answer is yes; however, it is even more important to consider the potential effects on the nucleic acid segment in the homeobox. In the Cambrian Sea, we cannot explain the explosion based on accelerated evolution in just one lineage. Remember that multiple different phyla were produced, and in each phylum many different new classes and orders of organisms. In most cases these different new types of organisms were widely different in appearance and body plan. This has been the puzzle. How could multiple related lineages demonstrate a phenomenal acceleration of evolutionary innovation and then just as suddenly settle down to more normal rates?

Shape and Adhesion in the Homeodomain

The homeobox domain (DNA sequence) was important because it revealed underlying similarities in development between mammals and insects. In addition, as its mechanism of operation was better understood, it provided evidence that the fundamental process of gene regulation (the turning on and off of genes) was similar between bacteria and multi-cellular animals. The homeobox encodes a DNA recognition sequence—that is, it forms a marker that sticks to a particular segment of DNA that it recognizes. Then it "calls in" a copying enzyme to start the transcription and translation process that will ultimately produce a working protein from the gene it has located.

There are cations known to be involved in the homeobox recognition mechanism, and they clearly involve electrostatic interactions as well. Not all of the details of the entire process are known. There are homeotic selector genes that are first transcribed into gene regulatory proteins that contain the DNA-binding *homeodomain* (a string of about 60 amino acid that do the recognition task). The homeodomain "recognizes" and sticks to a segment of DNA called the homeobox. The homeobox is a signal marker in the DNA that lets the cellular machinery know where to find the *Hox* genes for determining the body segments. Amazingly, the various homeoboxes and *Hox* genes that control the assembly of the full set of body segments are laid out in the same physical order along the chromosome as the order of the body segments they will be shaping (Alberts et al. 2002).

All of these processes—the sequence recognition by homeodomain regulators, the process of copying and replicating the rigidly complex organization of the *Hox* gene region, and the process of individual sequential selection along the series of slightly different *Hox* genes arrayed in their line—all of them are affected by electrostatic processes and are only partially understood. In other regulatory phenomena in mammalian DNA, there are architectural proteins involved that cause and shape the bends and loops in the DNA in order to achieve complex regulatory effects. The rigid spatial array of *Hox* genes suggests that architectural manipulation plays an important role in their function. In fact, it is very well documented that at least one metal cation-based protein is involved in homeotic gene function. This involves the pair of zinc cations in the in the LIM-homeodomain region (Sanchez-Garcia 1993). (LIM is an acronym for the three gene products beginning with the letters L, I, and M that led to the initial discovery of the LIM domain).

The experiment with HIV reverse transcriptase is important because it was the first time that it has been shown that a type of environmental change that took place at the dawn of the Cambrian era could have had a global effect on evolutionarily significant genetic processes, particularly as they affect body plans. The dramatic shifts in glaciation that took place at the Cambrian-Precambrian boundary 543 million years ago surely were capable of causing an electrolyte shift of significant magnitude (Corsetti 2003). Major shifts of the magnesium calcium ratio in the ancient seas are well-documented (Dickson 2002). Our experiment shows how environmental effects such as these can alter the fabric of evolution, particularly as it pertains to the genes that control axial HOMs.

Impact of Selection at the Level of the High Order Module

The impact of natural selection at the level of the organism is that it alters a process that produces random alteration in DNA so that, instead, it results in living organisms capable of reproduction, and species capable of changes over time that enhance their competitiveness for survival. You could say that natural selection both focuses and optimizes genetic change. We may infer that the more intensive the selective pressure, the more rapid the change.

The existence of multiple copies of axial HOMs within an organism offers several benefits for the organism. The tremendous success of the two most prolific bilaterian species generators—the insects and the vertebrates—attests to this. There are very few non-bilaterian animals with significant species presence in the world. Because segments can be duplicated and individually identified by the genetic mechanism, it is possible for major variations in one segment to proceed without altering an adjacent segment; for example, legs can be preserved while wings are added. Another major advantage is the tendency of HOMs to be capable of very high levels of pleiotropic change. This means that a single, small genetic change can result in extensive changes throughout the animal, affecting dozens of tissues of various shapes and functions. This very high level of pleiotropy is a double-edged sword: Small genetic changes affecting the *Hox* system are quite liable to spell disaster if a poorly functioning array of changes is made. On the other hand, it allows for rapid elaboration of a major structural reinvention of the organism. This is what happens when new body plans are generated as in the Cambrian explosion. It also accounts for events in which multiple minor changes appear throughout an organism, changes that alter its ecological niche and its relationships with its ancestral species. This is precisely the sort of change I am proposing for the origin of our lineage of upright hominoids.

7

How Vertebrae Came to Be

Introduction

The Darwinian Thresholds

The advent of evolution by descent with modification (Darwinian evolution) now appears to have commenced independently at least three times. In each case the onset was achieved approximately 1 billion years ago by a transition that can be described as "crossing the Darwinian Threshold." This term was proposed by Carl Woese, a microbiologist from the University of Illinois, in a landmark paper published in 2002 in the prestigious journal *Proceedings of National Academy of Science*. In a subsequent paper titled "A New Biology for a New Century," Woese (2004) details the history of biology that made this critical threshold virtually invisible to the scientists of the 19th and 20th centuries.

The events surrounding the chemical initiation of the processes that would ultimately lead to cells and genes remain clouded in uncertainty. Current thinking contends that they may have involved metal-carbon reactions on the edge of an ancient undersea volcanic vent (Wachtershauser 2006). Subsequently, for the next 2 to 3 billion years prior to the Threshold, the various components of modern biological systems—including the basic mechanics of translating nucleic acid codes into proteins—emerged in various biological entities. However, as you will remember from Chapter 1, the biological world was dominated by *horizontal gene transfer* (HGT), or the exchange of genes between unrelated types of organisms. Because of the dominance of HGT, new biological capabilities and enzyme systems that emerged gradually became distributed among numerous different types of organisms. Woese has shown that three major types of organism—the bacteria, the archaea, and the eukaryotes (plants and animals whose cells have a nucleus)—separated from each other at a time when there was still a considerable degree of HGT in the world (Woese 1977, 2004). Each of these three groups truly commenced existence as a separate lineage when a single individual organism crossed the Darwinian Threshold. Again, this is something that appears to have happened independently at least these three times.

What happens at the Threshold? The organism crosses the Threshold by establishing a relatively resilient barrier to incoming genes from the surrounding world. The organism will rely instead only on its own genes. Any new genes or any changes or improvements in the existing genes will now need to emerge, generation after generation, through modification of the existing group of genes that the organism already has. This is called *vertical gene transfer*, from parent to offspring without any sharing with other lineages. Once the Darwinian Threshold is crossed, evolution in its classic sense commences in that group. Certainly there may have been other lineages that started in this fashion and have not survived, but we do have the three major groups of bacteria, archaea, and eukaryotes.

The events of the eukaryote Darwinian Threshold are particularly staggering to comprehend. Unlike bacteria, the cells of plants and animals contain a number of elaborate internal structures (organelles), such as the mitochondria that burn oxygen to provide energy and the nucleus that organizes and manages the genes. It has long been thought that the original eukaryotic cell (the urkaryote; Woese 1977) may have incorporated some bacteria-like cells into its interior. Some of these organelles still hold their own separate sets of genes, and some of them seem to have mixed their genes in with the genes of the master cell and allowed all of them to be managed together in the nucleus. In any case, once the Darwinian Threshold has been crossed, it becomes possible to more reliably track the evolution of lineages and to accurately identify branch points and speciation events.

Out of the one-celled eukaryote creatures, a lineage emerged that experienced a most remarkable and sudden transformation—one that has made all subsequent animal life on earth possible. This was a relative of the single-celled amoeba and the slime mold that one day engulfed a cyanobacterium (Steenkamp et al. 2006). That individual cyanobacterium was the grand ancestor of all the plastids (chloroplasts and the plastids of red algae) that carry out photosynthesis, the capture of light to convert ammonia into protein, sugar, and oxygen (McFadden and van Dooren 2004). This occurred about 600 million years ago, leading to the blue-green algae, the red algae, and the plants. This is an example of *endosymbiosis*, in which one type of organism comes to exist inside of another. In this way, in a totally non-Darwinian fashion, the plants come into existence and create the oxygen-rich planet in which the animals came to evolve. This is yet another example of a modular change, in which Darwinian evolution acts only as a secondary player, engaging natural selection to act as an adjuster and polisher to fine tune and optimize. In this unquestionably very major event in the History of Life, a new type of organism suddenly comes into existence in a bio-ecological space that is utterly and spectacularly different from the world of its immediate parents.

The next major event was the capture or development of a fundamental complex that drives movement. This is the *flagellum* seen in the reproductive element (sperm) in all branches of fungi and animal species, and functioning in a variety of other tasks in various descendants. In fact, the evidence is now clear that the presence of a motile sperm is the defining common character of all fungus and animal species. Along with those groups—and including a number of single-celled organisms that share that capability—we have the new group name, the Opisthokonta (Cavalier-Smith 1998, Huang et al. 2005).

Many of the insights into the sequence of events in the origin and diversification of life on this planet have been clarified recently by decoding the genomes of various organisms. The method of encoding proteins as information in DNA is the most common shared phenomenon of all life. Nonetheless, despite the critical nature of this process, the mechanics and details of many aspects of this vary considerably from domain to domain, from kingdom to kingdom, and even among groups within the kingdoms. Nonetheless, there is enough similarity in the fine details of the process to make a single origin of it absolutely certain (Robinson and Bell 2005).

Some additional light is shed on the origins of our DNA and RNA processes by obtaining information from the viruses. A virus is a collection of DNA or RNA wrapped in a membrane incapable of reproducing itself without taking over the machinery of the host organism. As such, viruses are not considered independent forms of life. There are viruses that act only on bacteria (bacteriophages), viruses that act only on archaea (archaeoviruses), and viruses that act on eukaryotes (plants, animals, fungi, and related one-celled creatures). In fact, there are 10 times as many viruses on earth as all other living things put together. They have always been—and continue to be—important vehicles of horizontal gene transfer, in that they are able to capture genes from one organism and insert them into the genome of a comparatively unrelated organism.

Over the years, various theories of virus origin have been considered: Did they exist first before other life? Are they the remnant of a slimmed-down organism having shed most of its machinery? Or did they arise by out of a segment of the genome of an unfortunate parent life form? It is now increasingly clear that they arose out of organisms more ancient than the Last Universal Common Ancestor (LUCA), the ancestor of all existing forms of life (Forterre 2006, Koonin et al. 2006, Prangishvili et al. 2006). Despite the blurring effects of horizontal gene transfer before the arrival of the Darwinian

Thresholds, the viruses share an array of different mechanisms of processing DNA and RNA that show they arose before the LUCA settled on the existing machinery seen in all living creatures.

The Progress of the Cells

Following the emergence of the major lineages of nucleated cells with their various full complements of internal, endosymbiotic organelles (the chloroplast, nucleus, and mitochondria), the single-celled organisms made a transition into multicellular forms. This apparently happened independently three times in the plants, the fungi, and the animals. Huge cooperative assemblies of individual cells emerged, and came to demonstrate sub-specialization so that all were not identical. They shared copies of a single genome that spelled out their specializations and rules of cooperation. Out of the algae emerged the plants; out of the one-celled fungi emerged the more complex *eumycote* phyla that produce mushrooms; and finally, out of the one-celled animals emerged the multi-cellular animals, the metazoans.

The Advance of the Animals

The complex animals composed of many different cells include some very simple holdovers from the most ancient times, such as the sponges and the hydras (see Figure 7-1). However, starting with some very simple types of worms, and extending through the realms of insects, spiders, crustaceans, and on through the vertebrate animals, all of the creatures share the echoes of another dramatic advancement in a shared ancestor. This is the advent of a process we now call terminal addition (Hughes and Jacobs 2005, Jacobs et al. 2005). In terminal addition, the growing embryo elongates from the posterior end. For the most part, this involves the overt formation of segments or serially repeating structures. The details of the elongation process, and the degree to which actual segments can be observed, vary from phylum to phylum. But the genetic dissection of the process makes it clear that this entire remarkable group of animal phyla descend from a common ancestor in which *Hox* and *Pax* genes control the process of terminal addition.

Not only does this understanding provide a magnificent insight into the shared origins of so many apparently different types of creatures, from worms to humans, but it also greatly improves our understanding of the mechanisms by which all of these organisms are assembled. Furthermore, it seems to mark the great organismal innovation that sparked the Cambrian explosion and generated most of the existing animal life on this planet. It seems likely that all of these creatures either have repeating segments somewhere in their bodies, or have lost them in descent from ancestors that did have the segments. As Goethe suggested 200 years ago, and as studies of the *Hox* genes now confirm, the genetics behind the segments in all these creatures is somehow related.

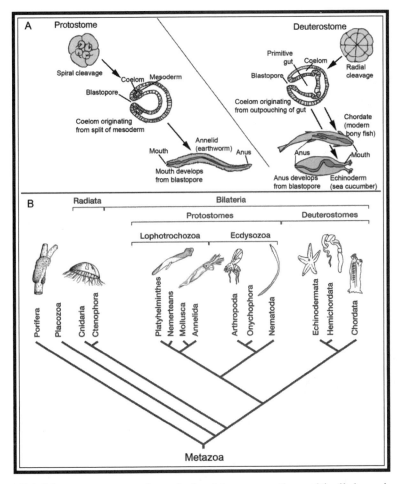

Figure 7-1: Metazoan systematics: relationships among the multicellular animals. The bilateria share bilateral (right/left) symmetry, in contrast with branching (sponges) or radial symmetry (jellyfish). The bilateria are divided into protostome and deuterostome groups based on which end of the embryo turns into the mouth. The vertebrates are deuterostome bilaterians, but our group includes some of the very simple animals shown in the diagram.

Many in the molecular developmental scientific community have argued (incorrectly) that segmentation is so different and carried out in so many different ways among the different groups in the bilateria that it must have evolved independently, and so what we see are similar bits of anatomy, but nothing fundamental. The reason for this confusion is that much of the work on segmentation started with fruit fly genetics, and we now know that these insects have extraordinarily unusual and modified segmentation that is a unique morphogenetic specialization.

In 2003, Angelika Stollewerk and colleagues, geneticists from the University of Cologne in Germany, reported in *Nature* that the detailed genetic mechanisms that produce somite segmentation in humans and in spiders appear to be astonishingly similar in the role of an array of genes. The similarities are so comprehensive that it is impossible that they occurred independently of each other. Not only are the homeotic *Hox* genes and *Pax* genes that direct regional specialization shared by all bilaterian animals, but the mechanics of segmentation (the embryologic formation of distinct repeating units along the main body axis) seen in the vertebrates appears to be close to the original bilaterian segmentation mechanism. This also bears heavily on how we view the various types of bilaterians—some phyla of worms, for example—that do not have any evident segmentation. Rather than proving that segmentation evolved independently in various groups, the data from the spiders tend to prove that unsegmented phyla have lost the segmentation mechanisms that were present in their ancestors.

The idea that all animals could be assembled from fundamental building blocks was the poetic science concept that led Goethe to his concept of the archetype. From 2003 to 2005, advances in molecular biology have shown that these serially repeating building blocks do exist, and that they can explain the construction and design of more than 99 percent of the animal species in existence. The same *Hox* genes that spell out the construction of the mammalian body form are also at play in the insects, the mollusks, and most of the worms of the world. The universal monomers that Goethe identified, and the proposal that Geoffroy Saint-Hilaire made regarding the universality of body segments among the animals, are now further proven correct.

The Matter of the Urbilaterian

Long ago, in the ancient sea of 550 million years ago, there lived a creature that was most remarkable in its future potential. The very existence of this strange ancestor was hardly even guessed at 15 years ago, but it has turned biology upside down in stunning ways. To an extent that no biologist could have imagined even 10 years ago, this odd animal has shown that Goethe's 18th-century philosophical theory of the animals seems to be a reality.

There is no real name for this creature, but biologists have begun to refer to it as the *urbilaterian* (pronounced oor-buy-lateer-ee-an). The "ur" means the ancient ancestor or the initial predecessor, and the term *bilaterian* refers to all the animals in nature that have bilateral symmetry (a right and left side that are more or less mirror images of each other) and a body that is constructed out of a series of segments. The term was suggested by two biological chemists at UCLA, Edward DeRobertis and Yoshiki Sasai, who published a very dramatic paper in *Nature* in 1996 that drew attention to the surprising way that modern molecular biology was revealing support for pre-Darwinian biological theories.

There is no fossil record to show that the urbilaterian existed. However, the echoes of the existence of this creature are recorded in stunning clarity in the genes of every living descendant, from fruit fly to spider to human. Certainly, it is difficult to fully imagine the appearance of an animal that can be a common ancestor of an ant, a worm, a spider, a fish, *and* a lion. It would be one thing if we were trying to imagine a simple blob whose later descendants would one day develop the characteristics of various types of advanced animals along several independent lines. However, this creature seems to have already possessed most of the advanced features, such as body segments, brain, nervous system, right-left symmetry, and so on. Most likely there will never be any fossil evidence of the urbilaterian. However, we know a great deal about it because of the shared *Hox* or homeobox genes we find scattered among its descendants. The presence of full sets of these genes in widely unrelated modern animals shows that nearly all of them were present in the ancient common ancestor.

All of this has to do with a common construction plan for assembling an animal. You may remember from biology class learning about the ball of cells that make up the early developing embryo. Somehow, that ball of cells starts to gradually take on the shape and features of adult creature. It elongates and develops a head end and a tail end; a right side and a left side; a head with eyes; arms and legs sticking out of the body; a means to eat and process food (mouth, stomach, and anus); and a nervous system to integrate and operate the whole arrangement. This seems to be a fairly basic list of items, and these features are exactly what we expect to find in the urbilaterian. However, although we find all these features in various descendant species as diverse as fruit flies and humans, there are also many descendants, such as clams and starfish, that do not have these features. This begins to hint at why the urbilaterian concept is so unsettling from the point of view of standard Darwinian evolutionary biology. This common ancestor appears to have a far more complex and advanced design than many of its descendants. In addition, we must also cope with the fact that there have been very few major new body elements that have come into existence since that very ancient time.

Gradually, geneticists have made progress in unraveling the signals and programs written out in the genes of animals that direct the embryo to mature from a ball of cells into something with all of the features we recognize as the parts of an adult animal. Out of a complex welter of odd-sounding gene names, a fundamental order has finally started to emerge. We are beginning to understand in detail how organisms are assembled, how the assembly instructions are written, and how these instructions are shared and modified among the animal phyla. The biggest surprise is that, no matter whether you're looking at a spider, a lobster, a lizard, or a human, the fundamental construction plan is written out in almost identical form. Granted, biologists had expected for many years that the construction plan of the vertebrate animals would have many common features. However, it was always thought that the invertebrates would have a completely different construction sequence. This was really the main thrust of biological science in the 19th and 20th centuries. Goethe, working from philosophy, suggested that there must be a fundamental subunit or building block out of which all animals are constructed, and that building block was the body segment exemplified by the vertebrae in vertebrate animals. The idea that this implied similarity between insect segments and the vertebrae in chordates was fully articulated by Étienne Geoffroy Saint-Hilaire in his 1822 publication for the Natural History Museum of Paris (see also Geoffroy Saint-Hilaire 1820). Since the 1830s, however, most biological scientists working from the point of view of scientific observation have rejected this concept. There is a very large body of scientific work showing that there is no similarity at all between the segments of a worm or insect and the vertebrae of a lion, a cheetah, or a human. However, this body of work now proves to be incorrect. Goethe and Geoffroy had it right in the first place.

The prevailing view was that among mammals, for instance, the body was not fundamentally segmented. Yes, the spine was made up of a series of vertebrae, but this did not appear to suggest anything fundamental about segments underlying the whole body plan. From eel to human, none of the vertebrate animals have the sort of distinct body segments that are seen among honeybees or inchworms. Even if a biologist were to accept that the repeating body elements in the vertebrates were "true segments," this did not seem to lead to any fundamental significance for this feature among all animals. This is because among 20 different phyla of animals, there are only three phyla—arthropods (insects), annelids (segmented worms), and chordates (vertebrate animals)—that have clear evidence of serially repeating structures in their bodies. Not only are segments essentially rare among animals, but the three phyla that have them are not obviously closely related to each other. For instance, the chordates are grouped with the phylum of the starfish, which is not segmented. The arthropods have traditionally been placed among six other phyla that share many features with them—but not segments. Annelid worms have been placed among 10 different phyla, none of which have segments.

Here we begin to question the interest in this peculiar issue. If segments are rare among animals, are we really overstating the importance of it because we ourselves happen to be in a group that seems to have segments? A more strictly Darwinian view casts the issue in the cold hard light of adaptation. Firstly, it can easily be pointed out that among mammals, vertebrae are not the only serially repeating structures. What about teeth? Although the teeth are fairly similar in shape in some vertebrates, the earliest vertebrates did not have teeth. Moreover, some mammals have a specific series of incisors, canines, premolars, and molars. Different groups of animals have different numbers and shapes for these teeth, but the top and bottom jaw (as well as the right and left side of the mouth) each have a tooth series. More interesting is the fact that we know that the embryology of the lower jaw (mandible) is quite distinct from that of the upper jaw (maxilla). Nonetheless, the upper and lower tooth rows are very similar (they usually match) and clearly change together in evolution. Then there are the digits in the hands and feet—five on each hand or foot, and three pieces (phalanges) in each digit. These are serially repeating structures as well. Even the arms and the legs share many features with each other. Once again, it is easy to see that the earliest vertebrate animals did not have arms, legs, hands, or feet. As with the teeth, these serially repeating structures appear to have evolved as adaptations. It's very possible that vertebrae might have a similar history.

The argument is therefore made that any bone, muscle, or skin structure can be fashioned out of serially repeating elements, but that this is in no way fundamental. We can look for the adaptive significance of the repeating structure from the point of view of natural selection, and the various specific advantages the serially repeating structure confers upon the organism. Annelid worms have segments because this helps them coordinate their inching movements. Insects have segments so that they can move their externally armored bodies. Chordates have vertebrae to provide a linearly rigid central skeleton for supporting movement while preserving the ability to bend. My own view here is quite different. I grant that in a vast array of animal forms, segmental designs present in the ancestor have been lost in the descendants. However, in the three animal phyla with the largest diversity of species (currently existing and across time), segmental designs have been maintained. Further, I believe that aspects of the body design that arise from the principal segmentation of the body are generally responsible for the most fundamental aspects of the function of the animal.

It is certainly quite clear that the cellular and genetic mechanisms that bring about segmentation are quite varied across the Bilateria (Minelli and Fusco 2004, Liu and Kaufman 2005) and even between different groups of insects (Arthur 2002, Arthur and Chipman 2005, Damen et al. 2005, Fusco 2005, Minelli and Fusco 2005). Nonetheless, the underlying genetics of terminal addition, as well as many of the mechanisms of producing segments in various distantly related bilaterians (for example, spiders and mammals) make a fundamental homology and similar role for the phenomenona of terminal addition and segmentation impressively clear. The findings of Stollewerk and colleagues (2003) settles the issue. Not only are the *Hox* and *Pax* genes that determine the identity or characteristics of body regions and segments shared among the Bilateria, but the original segmentation mechanism (*Notch* signaling) is extraordinarily ancient, and the version employed by vertebrates appears to be the original ancestral mechanism of the bilateria. In vertebrates, this fundamental role for the products of terminal addition (the somites and vertebrae) has been obscured, in part because many biologists have had difficulty understanding the fundamental ways in which vertebrae affect the design and behavior of mammalian and other vertebrate species. This chapter explores the history, assembly, and basic design of the vertebrae themselves.

Hox Genes and the Bilateria

The study of *Hox* genes has revealed some explanations for some of the great misunderstandings that have hampered our appreciation of the unity of serially assembled segmental construction in the organization of animal life on this planet. Because the fundamental similarity and ancient common origin of bilaterian segmentation was not proven until 2003, the confusion on this point in the previous decade is reflected in many textbooks and scientific articles on the subject. These misunderstandings are quite current; at the time this book went to print, recently published textbooks on comparative vertebrate anatomy have not yet been updated to reflect this kind of information.

The most common result of this misunderstanding is the emphasis on distinguishing the role of homeotic genes from the issue of segments. Modern morphogeneticists point out that *Hox* genes determine the regional specialization of tissues along the length of the body axis, just as *Pax* genes determine the specialization from front to back (ventral to dorsal). Just as the 19th-century scientists who rejected Geoffroy's position, they argue that segments are effectively rare locomotor specializations that turned up long after the role of the *Hox* genes was established. Thus, regionalization genes would be ancient but segmentation would not be. What is now clear is that regionalization determined by *Hox* genes may be preserved, even when overt segments are lost or suppressed. However, the two very successful lineages of the arthropods and the vertebrates not only employ segmentation, but they demonstrate the ancient or original mechanism. Granted, there are many organisms that lose or greatly modify genetic components of the segmentation system, just as *Hox* genes are lost in some groups. It is fascinating and informative to know that so many forms of life have existed for hundreds of millions of years after losing the genetic instructions for complex construction that the vertebrates share with the common bilaterian ancestor. However, it is correct to emphasize the fundamental nature of segmentation, and it is appropriate in this new light to accept the fundamental relationship between the series of discrete regionalization genes provided by the *Hox* system and the serially repeating segments whose construction and shape they direct.

The Right-Left Mirror in Bilaterian Origins

The epochal evolutionary event that has produced the grand and magnificent array of the bilaterian animals is based on mirrors and duplications. The emergence of a mirror or duplication event is never a simple matter of gradual Darwinian evolution, with adaptive natural selection acting upon variation in a population of animals. Consider the most fundamental aspect of the Bilateria: They are literally bilaterally symmetric. Humans included, they have more or less matched right and left sides. Before the emergence of the Bilateria, however, animals did not look that way. Imagine discovering a fungus or mushroom in which a complex shape on its right side was matched by an identical complex shape on its left. Or how about a shrub, or a tree with the left side a mirror image of the right? I believe it can be confidently proposed here that the emergence of right and left mirrored copies of the body plan in the grand ancestor of the Bilateria was a sudden and total event. Whether it preceded or followed the innovations of terminal addition and segmentation, these three features together are the fundamental aspects of the Cambrian explosion that produced so many of our existing phyla of animals. Indeed, mirror mutations are known to exist (Sagai 2004).

In a previous chapter, I pointed that the 180-degree flip that converted an early invertebrate into the first vertebrate organism must have been a sudden and abrupt event in a single individual's mutation. It was not a matter of evolution gradually turning the organism over at maybe 1 degree every 100,000 years through thousands of generations of slow gradual evolution. Rather, the flip happened as a single non-Darwinian event. In other words, it employed neither the mechanisms proposed originally by Darwin, nor the revised version expressed in the Modern Synthesis. This is not a matter of gradual shifts of gene frequency in a population that is driven by natural selection. In an instant, due to a mutation in a critical gene affecting embryological formation, a new type of animal—indeed an entirely new class—came abruptly into existence. This change was massively pleiomorphic, affecting many other morphogenetic processes and events taking place later in the drastically altered embryologic formation sequence. Fixation of the new trait was also instant, and no amount of gradual natural selection could slowly return the descendants of the vertebrates to recover the ancestral invertebrate anatomy.

Similar considerations apply to the origin of bilateral symmetry. It is not a matter of starting with a unitary ancestral organism with the radially symmetric design of a jellyfish, adding simple linear structure or the branching anatomy of the sponge, and then gradually developing a second copy, growing ever larger and budding from the side over many eons. What must have happened instead was a single morphogenetic event in which a single animal was born resembling a conjoined twin, with a duplicated, paired right and left side. Then, applying the mechanisms of chromosomal speciation as outlined in Chapter 4, the immediate descendants of this bizarre symmetrically paired mutant founded a new species. That species was the grand ancestor of all the bilateria.

Modularity Across the Midline

The right-left duplication of the organism was similar to the segmental duplications discussed previously. The duplication involved everything all at once: muscles, respiratory structures, and the nervous system. Before this event, it is clear that the most elemental nervous systems—those in the radially symmetric organisms such as the jellyfish, for example—were capable of directing signals in a linear direction from one neurally active cell to the next. And, as we know from the coordinated beating movements of the jellyfish, there were pacemakers in the nervous tissue. At a regular steady interval, a swimming beat emerges from the pacemaker and is distributed in a timed sequence along and through the body of the jellyfish. A coordinated flow of contraction is produced, resulting in forward movement. Some simple systems of worm locomotion work in a very similar fashion, but these take advantage of segmentation. Fluid is pumped out of one body segment and into the following body segment, which then hardens and swells, pushing the worm forward until the next segment takes over (see Figure 7-2A). Annelids are a phylum of segmented worms that move in this fashion. The flowing, progressing wave of swelling gradually drives the creature along its way.

In the bilateria, however, a different type of movement becomes possible. There is a distinction between the right and left sides of the animal. This is how the S-shaped undulation of bilaterian locomotion first emerges. With legs projecting from the wormlike segments of a centipede, the undulation pushes the animal forward by contracting on the left while relaxing on the right. A wave of right-left alternating movement flows down the length of the organism from head to tail. This is a reinvention of the most fundamental type of movement among the opisthokonta (fungi and animals), which all share reproduction involving a motile sperm with an undulating tail. With bilateral symmetry of the entire organism, right-left distinction in the net, and pacemakers of the nervous system, the entire, complex multi-cellular organism now moves in this undulating fashion. The waves of movement that pass down the right and left side of the body are in synchrony with each other (they travel at the same speed). However, they are usually "phase shifted"— that is, the one side is maximally contracting while the other side is maximally relaxing. A slight time difference in the maximum between right and left sides would be a small phase shift. When one side is at its maximum while the other side is at its minimum, we can say they are phase shifted by 180 degrees (see Figure 7-2B3).

Figure 7-2: Locomotor patterns in invertebrates.

A1) A peristaltic wave moves backward from segment to segment as the animal moves forward; segments maintain their total volume.

A2) A peristaltic wave moves forward as the animal moves forward; segments change their volume as fluid is pumped from one segment to the next.

B1) Limbs on *Artemia* (brine shrimp), with waves progressing in phase on right and left sides.

B2) Comb plates in overlapping shingle formation along the body of the jellyfish *Pleurobranchia.*

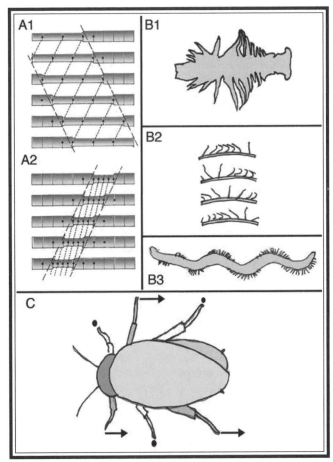

B3) A polychaete worm *Nereis* (phylum *Annelida*) can use its parapodia (feet) to walk or swim by mixing different types of metachronic waves.

C) Alternating tripod locomotion in an insect.

Locomotion With Legs

Reduction of the number of body segments involved in locomotion is another important trend among bilaterian animals that use a right-left, 180-degree, phase-shifted neural drive. Arachnids (spiders) and malacostracans (lobsters, shrimp) use four pairs of walking legs. Most insects (hexapods) use three pairs of legs, whereas vertebrate land animals use two pairs (or in bipedalism, one pair).

For much of the past 200 years it has been taken as a matter of faith that locomotion in invertebrates (worms, insects, spiders) is of specialty interest and has no direct bearing on how our own locomotion evolved. That underlying assumption must now be discarded. It is now clear that the vertebrate body and essential mechanisms of vertebrate locomotion developed together with the body organization and locomotor function found in many invertebrates, such as the arthropods. There was no separate pre-vertebrate ancestor that was some sort of unsegmented, hollow worm. Because we now understand that the occurrence and formation of bilaterally symmetric body segments in spiders and in vertebrates has a common origin to a very high level of genetic detail, it follows that our neurologic basis of locomotion also has a common origin. Of course, it is not just spiders, but most bilaterians that share this common neurologic basis of movement. Many of the various phyla of bilaterians have lost their segments, become much simpler in body form, or emerged into far more complex and specialized forms. However, the evidence of the morphogenes is clear: There must have been a common origin for the neurologic basis of locomotion in the bilateria.

Uneven Distribution of Advanced Body Designs

One of the great puzzles for the student of zoology in the past is the strange distribution of advanced anatomy across nature's species. On the one hand, we have advanced vertebrates with paired eyes, multi-segmented body structures, jointed limbs, a brain, nerves, muscles, skeletons, and complex food-seeking, social, and behavioral organization. When we trace the vertebrate ancestry back to the early chordates, we see some sessile, globular little animals that spend their lives attached to a surface, and have no skeleton, no muscle, no real brain, certainly no eyes, and no behavior to speak of, other than pushing water through a filter to try to collect food (see Figure 7-1). Then, elsewhere in the animal kingdom among the arthropods, we see insects, spiders, and crustaceans with eyes, skeletons, limbs, brains, nerves, muscles, and body segments, many with elaborate and complex behavior to obtain food, and communication to organize the behavior of large numbers of individuals. It has been assumed that these capabilities have developed independently twice, once among the vertebrates with our diminutive ancestors, and then quite independently among the arthropods. This understanding of separate evolution of the vertebrates and the invertebrates has been a central tenet of biology for nearly 150 years. However, this tenet appears to be incorrect.

Simple Forms Evolve From Complex Ones by Losing *Hox* Genes

A quick look at a comparative layout of the *Hox* genes shows the horrible truth (see Figure 7-3). The simple little ancestors of the vertebrates are degenerate forms that have lost many of the *Hox* genes that the modern vertebrates share with the fruit fly. The genetic instruction system of the fruit fly is much closer to our own genetic instruction system than the one employed by the simple ancestral chordates and hemichordates.

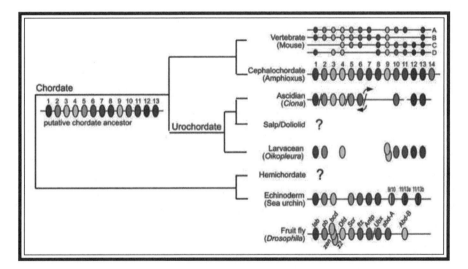

Figure 7-3: *Hox* genes in the deuterostomes. *Hox* genes that are present in both modern vertebrates and in insects may be missing in some chordates and other deuterostomes. Before this was understood, zoologists believed that complex animals, such as the insects and the vertebrates, evolved separately from simple ancestors. It now appears that there was a complex common ancestor and that some simple types of animals evolved by loss of genes and progressive simplification.

Let's be absolutely clear on this. It is not a matter of being more related or more evolutionarily close to arthropods as family members or descendents in a cladistic tree; that issue of sequential relationship has been all that is holy and important to most zoologists for many decades. Rather, it is a matter of sharing a full set of *Hox* genes with our more complex distant relatives than our very simple close relatives. This leads to the fascinating question of just how complex and developed the creatures were that were the common ancestors to the Bilateria. Did these ancient creatures have all of the body parts of a fly or modern vertebrate? Are all the simpler forms—the worms, the urochordates, the oysters—just degenerate animal forms that lost some of their *Hox* genes? Apparently the answer to all of these questions is yes. This conclusion is so different from what biologists have believed all their lives that it is akin to telling them that their religion is wrong. It produces outrage. Given the same set of information, it might have been possible to argue that *Hox* genes or no *Hox* genes, the original ancestors were simple creatures, but the complex forms developed later independently.

Bilaterians That Are Flipped and Inverted

However, an ineluctably clear and definitive review paper published in 2005 by two zoologists from the University of Basel, Lichtneckert and Reichert, in the major journal *Heredity*, really makes the model of an advanced complex common ancestor utterly unavoidable. This paper addresses—and I believe finally settles—one of the longstanding fundamental arguments against unity of the design of complex organisms.

Although insects have a complex central nervous system with a brain, the insect central nervous system is on the underside of the body (below the digestive tract); in vertebrates, the nervous system is on the backside of the body (behind the digestive tract). Also, the older name for these two categories of animals captures a related major difference. The arthropods come from the protostome group, in which the mouth forms from the back end (blastopore) of the embryo and the anus forms from the other end. The vertebrates come from the deuterostome group, in which the opposite situation holds true (see Figure 7-1). Because of this difference, it has been argued that the existence of a complex common ancestor for the insects and the vertebrates is highly improbable. To make a vertebrate embryo out of the starting form of an ancestor with an insect-type embryo, nature would have had to reverse the direction of its digestive tract, flip its body organization from front to back, and turn its body plan upside down to get the neural tube into the new location. This seemed to involve an improbable series of gradually evolving adaptations. Therefore, it has been argued that the common ancestor was so simple that chance took its two lines of descendants in opposite directions, most likely evolving a mouth and anus and a nervous system twice independently.

However, with the understanding of the *Hox* genes and the genetic mechanisms for laying out the main directions of the body axes during embryogenesis, it is now apparent that a simple and ordinary change in the gene function could flip all the axes in one sudden step. In order to prove this, we need to show that some structure, such as the central nervous system, is specified in detail in exactly the same way by the same genes. This would make it easier to accept that the central nervous system was already significantly developed and quite complex in a common bilaterian ancestor before the protostome/deuterostome flip took place.

Shared Genetics of the Vertebrate and Insect Brain

In fact, this is what Lichtneckert and Reichert demonstrated in detail in their article. Both the insect and the vertebrate brain have a "tripartite" ground plan—forebrain, midbrain, and hindbrain—and in both of these very distinctly related bilaterians, the *Hox* genes and a number of other quite independent gene families play the same roles in organizing the layout of the brain. This is an excellent example of how understanding the method of construction of an organism may do a better job of explaining the similarity among animals than what can be accomplished by understanding only their relatedness.

Regression and Conservation in the Fabric of Animal Life

When you look across the vast array of the tens of thousands of bilaterian species across 500 million years, you see that much of the tree is filled by species that have lost major complex assembly genes. This stands in direct opposition to any concept of progressive evolutionary development as a consequence of adaptation to the environment. It seems to show instead that the elaborate design of advanced animals with eyes, limbs, complex behavior, segmentation, a complex brain, and muscles to move with is really not necessary for successful life. Descendents of an ancestor whose genetic accidents robbed them of all these features are still alive and reproducing after hundreds of millions of years. The only counterpoint to this regressive explanation of animal life seems to come from humans. With enough complexity—as we seem to have—one species can come to dominate and commandeer all the surface resources of the planet and beyond.

Overall, my point here (which will be reiterated in greater detail later in the book) is that the fabric of morphologic change in life is not at all what we have been taught, whoever our teachers may have been. Classic evolutionary thinking shows a steady adaptive progress by which animal species become improved and more successful versions of their ancestor species. Modern evolutionary thinking shows a random progress to forms that are different just because they are different, and that survive because the changes in their form just don't make much difference. In fact, it seems that once a complex working design is established in nature, it has tremendous staying power.

After all, more than 99 percent of living animal species are bilateral, *Hox* gene organisms whose ancestors had all the genetics to create a three-part brain and virtually all of the other body systems. However, the existence of an advanced form does not prevent the success of less advanced forms that effectively descend from an advanced one. Evolution is constantly at work producing less complex animals that do just fine without the capabilities they have lost. In the Bilateria at least, creatures with the ancient primitive forms can be the most advanced.

From this point of view, nature long ago assembled a complex animal with a full complement of *Hox* genes and all the components needed for complex animal life. Turn this creature loose half a billion years ago, set up the principles of evolutionary change to act upon it, and, lo and behold, you steadily produce all the diverse animals of the modern world. This is very close to the archetype concept proposed by Goethe, and embraced by both the scientific and theological community in the first half of the 19th century. It is the idea that was laid to rest by Darwin.

The Emergence of the Somites

Just as the *Hox* genes are the genetic embodiment of Goethe's concept, there is also a literal muscle and bone embodiment in the developing embryo called the somite (see Figure 7-4). A somite appears as bumps near the back of a vertebrate embryo, one on the left and one on the right. Another pair of these somite bumps rapidly appears behind the first pair, then another pair, and so on, until there is a long line of dozens of paired bumps stretching from the head into the tail. If you ever watch a time lapse video of a chick embryo developing, you will see these segments forming one after another, in an ancient pattern repeated in every embryo of every chordate ever born across hundreds of millions of years. Very similar events take place in every insect embryo, and in fact in every embryo of most of the animals on the planet.

In vertebrate animals, the cells of each somite start to specialize and then go on to give rise to the cartilage, the muscle, and the bone that will make up that segment of the forming animal. The somites have a specific numbering related to the information in the numbered *Hox* genes, and have corresponding specific developmental roles or "fates" (see Figure 7-5). Although nearly all of them will produce cartilage, skin, muscle, and bone, the final shape of the tissue they form—a segment in the neck, part of an arm, part of a leg, or part of a tail—is determined by their position in the body sequence and is under the control of the *Hox* genes.

The somites can be matched, number for number, with individual spinal segments. Further, there are specific links between the number of somites and the *Hox* genes that direct the formation of local morphology. For instance, the boundary between the cervical and lumbar region reflects a boundary in Hox gene expression, although there is no one-to-one matching of *Hox* genes to somites. That is, once seven cervical somites have been elaborated, a different mix of *Hox* genes comes into play for the 12th somite, directing it to form into the first thoracic vertebra (see Figure 7-5).

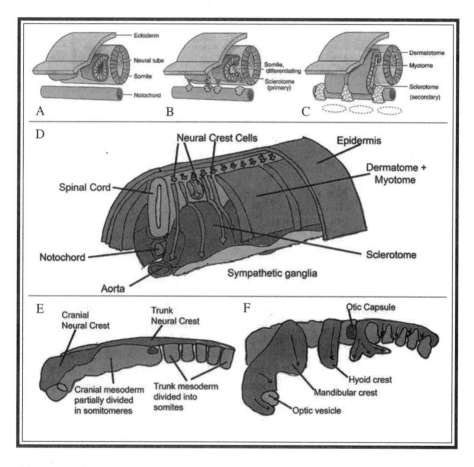

Figure 7-4: Somites, somitomeres, and neural crest cells have tissue identities determined as serially repeating structures by *Hox* genes.
A) Body segments in vertebrate animals form as clumps of tissue called somites. B) Somites develop into various types of tissue including the bony vertebrae and C) the muscles of that body segment. D) Neural crest cells in serially repeating groups migrate within a segment to reach their destined positions to form nerve ganglia and other tissues. E) and F) The head forms from segmented tissues; some have been called somitomeres. A/B/C from Kardong 2005. Copyright 2005 by The McGraw-Hill Companies, Inc.

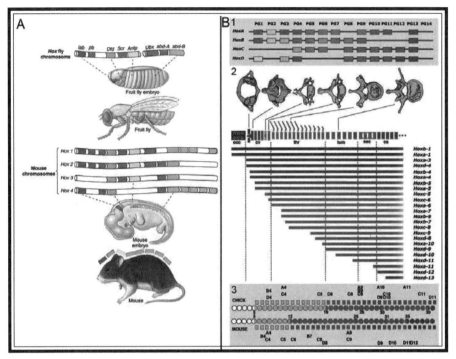

Figure 7-5: *Hox* specification of axial High Order Modules.
A) B1) and B2) The linear arrangement of *Hox* genes along the chromosome is maintained across most of the existing types of animals. In most vertebrates, there are four slightly different copies of the *Hox* gene array. From Raven 2005. Copyright 2005 by The McGraw-Hill Companies, Inc.
B3) Comparing birds and mammals, a shift in the expression boundaries is seen so that the cervical (neck) region and sacral region in birds are elongated, but the thoracic and lumbar regions are greatly shortened. The cervico-thoracic boundary is at somite number 19 in birds, but at number 12 in mammals.

Genetic clocks for assembling body parts

So how do genes actually carve out a series of segmental somite blobs? We now understand the genetic mechanism that is used. The system involves two chemical tricks: One is a genetic gradient, and the other is a cycling genetic clock. In an early embryo, a *gradient* describes a chemical arrangement in which, for instance, one end of the embryo has a high concentration of a signaling chemical and the other end of the embryo has a low concentration. The concentration decreases gradually as you proceed from one end of the embryo to the other. There are also genes that have now been discovered that can produce a repeatedly varying or oscillating signal. This is what makes up the genetic clock.

The formation point of the embryo gradually proceeds from the head towards the tail along a morphogenetic (body-forming) gradient. As the cells that will form the somites are being produced, the clock signal oscillates; when it is strong, somitic cells are formed, and when it is weak, the linear progress along the body gradient continues, but the production of the somite cells is cut off briefly. Then the signal increases again and the next somite begins to form. In this fashion, the developing embryo produces a series of separate lumps of tissue that underlie the segments of the body. There is a large and growing list of genes that participate in controlling and directing the process of segment formation, including some early discoveries such as the genes called *engrailed*, *Notch*, *Delta*, and *eve*. The situation is somewhat similar to the mechanisms of DNA replication described at the beginning of this chapter. We have an absolutely fundamental process with an inescapably common origin, but the wide array of existing species seem to have developed surprisingly diverse ways to get the job done. Alessandro Minelli and Giuseppe Fusco, two biologists from the University of Padova in Italy, have recently reviewed this array of segmentation mechanisms in helpful detail in an excellent 2004 overview.

Although somites per se are seen only in vertebrates, it is now also known that the genes that produce them are found in arthropods and have a somewhat similar function, although some critical parts of the process seemed to have been disrupted through genetic losses. It is therefore apparent that the somite-forming mechanism, and most likely the morphologic equivalent of the somites, were present in the ancient urbilaterian. Somites may embody the primitive or ancestral method of body formation in the Bilateria.

Segments for the head

As visually striking as they are, somites are not the only tissues that form this way. There are a few other varieties of serially formed tissues that are similar to their anterior and posterior neighbors, but are uniquely defined by their position along the body axis. This includes the blobs of *neural crest cells* that can form into tissues as varied as the jaw and the dentine of teeth in the head, while also forming nerve ganglia along the body. From the point of view of the history of science, the most interesting of all the segmentally organized tissues are the somitomeres. These are segmental "blobs" that are quite similar in their composition and fate to the somites. They differ primarily in that they are more difficult to see on the exterior of the embryo. Because of this, they eluded the observation of embryologists for nearly 200 years, and were not finally recognized until the 1980s. Despite the documentation of somitomeres by Meier (1979), and then Tam and Meier (1982), other embryologists have been concerned that somitomeres are not as distinct as somites (Noden and

Trainor 2005). However, evidence continues to accumulate that makes the existence of somitomeres increasingly certain (Kuratani 2005a, 2005b; Yamane 2005; Beaster-Jones et al. 2006).

Goethe proposed that the entire skeleton—including the skull—could be derived from vertebral segments. The failure of embryologists to observe the cranial somites was the major basis for rejecting Goethe's vertebral theory of the skull—one of the key points of his grand explanation of animal design. However, here again, time has proven that Goethe was correct. The discovery of the somitomeres showed that the same process that formed the vertebrae was involved in formation of the skull. Even more recently, as the genetics of the construction plan of the head and skull of vertebrates has come to be known, the concept of the "unity of design" mechanism for the segments of the body and the segments of the head has been proven.

The correct description of events is now increasingly clear in fine detail. There are seven somitomeres contributing to the segmental origin of the head. Segmentally organized neural crest cells in the cranial region help form the membrane bones that make up much of the skull and jaw. The somitomeres provide the cells that turn into the muscles for chewing and the muscles that move the eyes. The somitomeres are then followed by the cranial somites, which provide the cells that develop into the muscles for the tongue and throat. Finally, the first three of the main body somites typically contribute to the occipital part of the skull.

Conclusion—A Revisionist History of the Bilateria

Similar to a traditional story told and retold by a primitive tribe of people, or a canon of religion embraced and followed over the centuries, there is an endlessly repeated and increasingly embellished classical tale of how vertebrae came to be. Hard body parts appear as armor on early chordates, shifting into the enamel of teeth, but bone arises as the undersurface. Somites develop cartilage supports, which become bony, and the ancient fluid-filled notochord of the founding chordates becomes sheathed by a series of hard vertebrae, and so on. That story now needs to be reduced to a historical footnote.

If the genes of the somite segmenting mechanism are shared between spiders and vertebrates, then the somite must be an ancient common feature of the Bilateria. The various types of non-segmented animal phyla and the wide variety of different mechanisms of segmentation seen among animals reflect losses as well as diversifications. The vertebrates still form according to the most ancient segmentation system of the bilateria. In fact, we may even consider that bilaterians with somites and a deuterostome configuration were the ancestral type. The invertebrates may then be seen as descendants of the ancestral vertebrates by various flips of the body axis.

Homeotic Evolution in the Land Vertebrates

Homeotic Genes and Regions of the Spine

The essence of what the bilaterian animals have achieved in their various architectures is captured in the concept of regionalization by homeotic genes. The process of terminal addition adds length to the body, and the mechanics of segmentation allows those segments to be independent modules. The homeotic genes tell each segment what its individual identity will be, relative to the segments before and after it along the length of the body.

In the arthropods (insects, spiders, crustaceans), the impact of the homeotic genes is dramatic to observe. A series of segments has identical legs, and then another series has an entirely different shape. Large classes of these animals share characteristics that can be described in terms of which segment does what. In the aquatic vertebrates in the fish group, however, the segmental identities seem to have a far more subtle impact; there is a subtle steady gradation from the front of the fish to the tail that results in a slow decrease in overall size from head to tail. Although there are standard locations for fins along the body length, the tissues that strongly display segmental architecture (vertebrae and segmental sheets of muscle) have only a gradual pattern of relative change.

In the emergence of the land tetrapods we see a major transformation of this vertebrate character. A comparison between the latest common ancestors of the fleshy-finned fishes (sarcopterygians) and the first tetrapods shows stunning advance in the degree of regionalization. This is clearly seen in such ancient forms as *Ichthyostega* (see Figure 8-1), which is among the most ancient of all known tetrapods. It marks the emergence of land vertebrates from the fish in the Devonian era of more than 400 million years ago.

For many decades, lack of interest, insufficient knowledge, and little attention to the spine led to reconstruction drawings of this animal that suggested a relatively undifferentiated spine, similar to what is found almost universally in fish. Creating these kinds of vague drawings of spinal structures is actually a recurrent problem in the paleontological literature. Reevaluation of *Ichthyostega* now reveals distinct cervical, thoracic, lumbar, sacral, and caudal regions (Ahlberg 2005). This important recent discovery has had little impact on theories of vertebrate evolution because it has been viewed by many as merely a few specialist details about some ancient vertebral bones in one species. This is exactly how problems such as this arose and have been maintained over decades. However, this change in reconstruction reveals what is among the most epochal and fundamental transformations of the vertebrate body plan—namely, a transformation that set the stage for much of what would come to follow.

The appearance of the five major regions of the spine at the very dawn of the history of the land tetrapods shows that the homeotic terrain had experienced a figurative massive earthquake. Distinct regionalization of the body axis provided the basis for generating an array of new body plans to deal with the new realities of life on land. In addition, the very same homeotic genes that controlled regionalization in the spine also controlled regionalization in the long axis of each of the limbs, as well as in the transverse axis of the hands and feet.

In fish, there are fins with a simple organization; in tetrapods there are five components to the long axis of each limb. For the arm these are: 1.) shoulder/ pelvis (limb girdle); 2.) humerus/femur (proximal limb bone); 3.) radius, ulna/ tibia, and fibula (distal limb bone); 4.) wrist/ankle; and 5.) fingers/toes (digits). Of course, in the transverse axis of the digits there are five fingers and five toes. Five regions in the spine, five parts of the limbs, and five digits—all controlled by exactly the same homeotic regionalization genes. You can read any number of textbooks on evolution or vertebrate paleontology without seeing a hint of this homeotic perspective on the origin of the land vertebrates (tetrapods).

Based on a Darwinian perspective, we expect to see gradual transformation of the fish body over eons as trunk and fin body parts are gradually reshaped to help the ancestral tetrapods conquer the land. However, let's consider a remarkable alternative that has never been fully explored: What if the fundamental event in the origin of land vertebrates is the onset of a five-component regionalization capability in the homeotic genes? Suddenly and in a single event, the uniform spine of the fish is converted into the cervical, thoracic, lumbar, sacral, and caudal regions of the land vertebrates. The same revolutionary, five-part linear homeotic series is also applied to generate the five parts of the limbs and the five digits. This means that a single change that led to the five-part homeotic specification simultaneously produced the spine of land vertebrates, the limbs of land vertebrates, and the hands and feet of land vertebrates. This is actually what seems to happen in the fossil record.

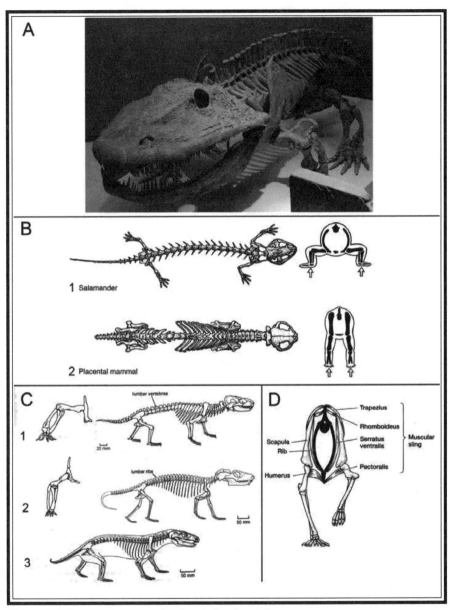

Figure 8-1: Locomotor organization in land vertebrates.
A) At a very early point in the emergence onto land, fins became five-fingered digits, and the spine demonstrated its five regions.
B) C) D) Walking in amphibians and some reptiles involves fish-like body undulation, with the limbs far out to the side. In the evolution of mammals, the limbs shift to a new position below the body. B/D from Kardong 2005, pgs. 345 and 390. Copyright 2005 by The McGraw-Hill Companies, Inc.

To add to the impact of this is the fact that *Hox* genes also control the branching and ramifications that provide the greatly increased surface area of the lung in land vertebrates as compared to fish (Volpe 2000). This means that the same regionalization change that produced the spine and limbs of the first land vertebrates could have increased the oxygen transfer capability of the first land vertebrates. In the water, respiration took place at the head, digestion and reproduction took place in the long belly, and the muscles and bones of body had a single simple shared task of driving the animal along through the water. Now, however, the muscles and bones of the body had a major new task: Ventilation—the extraction of oxygen for respiration of the tissues—had to be accomplished by the body at the same time that locomotion was taking place. The head became a simple conduit for the air, and the gills ceased to function and simply withered away.

History of the Architecture of the Vertebrae

The Ancient Origin of the Vertebra

Now we get into the thorny issue of exactly where and when the real original tissues of the actual vertebrae first began to appear. There is no question that muscles occur in fairly similar form across the great array of animal species. For instance, there is a lot of similarity between the structure of insect muscle and mammalian muscle. In every creature with muscles, there has to be something for muscle to attach to in order to get the most out its potential for doing work. I have already made my argument that a series of sheets of muscles—paired right and left, with each pair lined up in sequence along the back of the animal—were an ancient feature of the urbilaterian. This is a consequence of our evidence for the very ancient history of the somites. Essentially, if these creatures had a series of muscle segments and a nervous system, and if they used them for movement, then there must have been something for the muscle sheets to attach to in each segment. This certainly was not a predecessor of the notochord, because there aren't any animals in which muscles actually attach to the notochord itself. Instead, it would have been some sort of non-elastic connective tissue for the muscle attachments. This had to be the original antecedent of the vertebrae.

I grant that I am now treading on highly controversial ground. This is a proposal for the existence of what amount to vertebrae in the common ancestor of the insects and the vertebrates. I think the latest evidence requires that this conclusion be not only considered but seriously respected as a valid hypothesis about a fundamental aspect of the History of Life on this planet as it pertains to the animals. This is also an intriguing proposal because it brings Goethe's idea to full circle. He thought that vertebrae were the common design feature, and I have pointed out that the more acceptable modern version involves *Hox* genes, or even somites and somitomeres. However, if the urbilaterian actually had vertebrae, then Goethe was literally correct in his original concept. We can describe these original monomers as "*Notch* generated, *Hox* determined, and *Pax* organized" serially repeating elements (NHP SREs, for short). Somites and vertebrae are therefore "NHP SREs."

What do we see in the more primitive fish? Most of these have either cartilage or bone around the notochord, arranged in a way that achieves the effect of a series of vertebrae. In sharks, the outer sheath surrounding the notochord is partly made of the connective tissue of cartilage, and this assumes the shape of vertebrae. The cartilage of each vertebral element forms into a roof over the spinal cord, the ribs attach to the vertebrae, and the sheets of body muscle act on the vertebrae to swing the shark's tail to make it swim. The design is similar in most other fish except that it includes bone as well as cartilage. In the embryo, the notochord forms independently of the forming vertebrae, and it does not come from the somites. The developing vertebrae actually surround, cut, and squeeze the notochordal cells in the various designs found among the vertebrate animals.

In amniote animals, such as reptiles and mammals, the notochord is always sectioned into pieces by the developing vertebrae. Clumps of notochord cells are found inside the disks between the vertebrae, where they continue to serve a shock absorbing or elastic function. However, when a human gets a ruptured or herniated disk, what is happening is that a lump of notochordal material has burst out of the disk to pinch a nerve. The outer ring of the disk is made from the front end of each somite, whereas the main parts of the vertebral bone are made from the following end (caudal half) of the somite.

Vertebrae in Three Pieces

Based on what we see in our own skeletons, and in fact what we see in most living animals, we think of a vertebra as a single bone. However, an examination of fossil vertebrae from a wide variety of ancient amphibians and reptiles shows that the modern vertebral bone has been formed out of three original components. One part is the roof, or *neural arch*, over the spinal cord. This part forms separately in the embryo and then fuses together with the main body of the vertebral bone. However, given the fact that a single solid vertebra is found in most living animals, including the fish, I suspect that the ancient land animals with a separate neural arch were just regressing a bit towards the embryonic design. The idea of a progressive evolution towards a single, solid fused vertebra doesn't seem to fit with the wide occurrence of a single, solid vertebra in fish.

The other issue that plays out across time is that there are two versions of the main body, or *centrum*, of the vertebral bone: the *intercentrum* and the *pleurocentrum*. The intercentrum tended to be a ring around the notochord to which the main head of the rib would attach. Various lineages of early land animals can sometimes be characterized by the organization of their vertebral bodies. In some, the intercentrum and pleurocentrum are equal in size (called *embolomeres*); in others, the intercentrum becomes the dominant element and the pleurocentrum is small or absent (*stereospondlyous labyrinthodonts*). In the lineage that leads to the reptiles, birds, and mammals, the intercentrum gradually disappears. In most mammals, the main rib head loses its attachment on the intercentrum, and ends up attaching partly to the vertebra in front of it and partly to the vertebra behind it. The pleurocentrum then becomes the dominant part of the vertebra. In most reptiles, the main rib head just moves onto the middle of the pleurocentrum.

The Spine in Vertebrate Locomotion

Swimming With the Spine—The Basic Design of the Fish

In the water, the requirements of the vertebrate spine are very different from those on land. Even if limbs are used in the water, they do not twist. However, propulsion will still cause some sort of bending. In fish, the bending is almost universally from side to side. In aquatic mammals such as the whales or dolphins, the movement has been reconfigured by the ancestral land animals so that the bending during swimming is up and down rather than side to side. In movement without limbs, whether in a fish or a snake, the spine plays a more direct and central role in driving locomotion. Although there are a few primitive chordates that use jets of water to push along a body that is stiffened by a non-bending notochord, animals with vertebrae tend to employ undulating movement to achieve locomotion.

In fish, the hallmark of the ability to move by undulation is a muscular system organized as a series of segmental stripes of muscle. The muscles in each body segment act on the ribs and membrane-like septations to bend the spine. In mammals, there are still some segmental muscles, such as the intercostal muscles between the ribs; but in fish, most of the muscle in the body is part of a segmented system. The fish moves by wavelike contractions that propagate down the length of the body and that are coordinated by the central pattern generator. At any given segment or region, the left and right sides of the animal alternate in their contraction and relaxation. Swimming in the aquatic mammals (whales and dolphins) is roughly similar except that the waves of contraction affect mostly the tail, and the tail portion moves up and down rather than side to side. Swimming by undulation is possible in fish because of the segmental motor system and the behavior of the spine. It allows bending, but in the power stroke it acts as a stiffener to push against the water and drive the body of the fish forward.

The Slither of the Snake

At one extreme for spinal stiffness among vertebrate land animals are the snakes. Although there is a single basic mechanical design—a long, flexible spine that engages in lateral undulation—snakes can move in a variety of ways. The simplest method of progress is simple "sinusoidal" progression, in which a wave of curving and countercurving motion progresses segmentally along the length of the body. As with swimming fish, a snake generally depends upon a high degree of sequential segmental coordination in its muscle firing in order to achieve locomotion. For most snake locomotion, the curving motion includes a force directed backward that allows the snake to progress forward. Sidewinding uses a different application of forces during undulation, which allows the snake to move perpendicular to its long axis rather than parallel to it. Snakes generally cannot use "inching" to progress along; this would require a vertical bend in the body rather than the usual horizontal bend. Snake vertebrae generally permit vertical bending only in the neck. If you assemble the loose vertebrae of the skeleton of a snake and hold one end of the spine, the remainder of the spine will extend straight out without flexing downward, although it will very easily bend laterally.

The Emergence of Vertebrate Walking

The predecessors of land vertebrates included the fleshy-finned fish, of which the most famous living example is the coelacanth. These fish have muscles extending into their fins as opposed to the usual bone, cartilage, and skin design found in the fins of the vastly more common ray-finned fish.

In early land vertebrates, or *tetrapods*, the limbs functioned quite differently from those of mammals (see Figure 8-1). One important difference is that the first part of the limb typically points out to the side, extending laterally away from the body. The second part of the limb, past the knee and elbow, then turns down to meet the ground. You can simulate the motion if you put your upper arms straight out to the side and bend your elbows so your hands are in front of you. The upper arm rotates around its long axis, pushing the hand backward. This is quite different from the mammalian limb movement, which you can simulate by fully extending your arm straight out in front of you. You start by reaching upward, as though you're reaching up to wave to someone, and then bring the whole arm downward.

One interesting feature of the primitive walking stance is that the animal can actually advance the lower limb by continuing with a fish-like undulation of the spine. With the right hind limb planted on the ground (see Figure 8-1), an undulation turns the pelvis along with the lower spine. If the right hip joint is fixed in position, an undulation will actually swing the left hip joint forward. Then, when the left foot is put down, it will have advanced in position relative to where the right foot was placed.

Walking that involves planting the limbs and applying undulation to advance the body is actually very similar to swimming in terms of the central nervous system. This recalls the issue of the motor pattern generator mentioned earlier in this chapter. Raising and lowering the limbs may be only slightly different from what the coelacanth did with its fins to paddle along through the water. When the limbs are used in this way, the same undulatory motion that drives the fish through water will advance the animal over land.

Mechanics of Locomotion in the Mammalian Spine

It is in the reinvention of walking that the mammalian body design is most transformed. The limbs move under the body, and the spinal motion pattern is profoundly altered. Although lateral undulation seems to be relatively suppressed in the spine when a mammal is walking or running, the exact motion pattern of the typical mammalian spine proved to be difficult to understand until quite recently. This uncertainty arose in part because of some long-standing, incorrect assumptions about how the spine functioned as a structural element in a mammal. The obvious analogy is a bridge, but this actually led scientists away from a correct understanding.

In the bridge analogy, the limbs are the vertical elements standing on the ground, with the bridge (spine) suspended somehow between them. One version of this was to look at the spine as a suspension bridge, with the rigid longitudinal element of the spine suspended like a roadway from the limb girdles, and the mass of the body hung below the spine. Another version pictured the spine as an arched bow, with the abdominal muscles acting as the bow string, and the entire unit mounted on the limb girdles.

However, the spine is not really all that similar to a bridge, mostly because it is constantly changing its shape and stiffness during walking. The spine does this under the power of its own musculature interacting with a complex set of struts and elastic ligaments. We don't have any bridges that work that way. (The only enduring image we have of a twisting, moving bridge is from the famous footage of the Tacoma Narrows suspension bridge as it resonated in the wind and collapsed spectacularly on November 7, 1940.) Based on the success of the mammalian spine, I would therefore conclude that the spine doesn't have very much to do with what we understand about bridges.

Complex Movements Seen by Cineradiography

An external view of an animal provides only a very limited sense of what the vertebrae are actually doing. In 1980, I did the first fundamental research into this problem by using a cine X-ray camera to film an opossum walking on a treadmill, with a simultaneous electromyographic recording of the back muscles. More detailed analyses have been published since then (Ritter 2001, Schilling 2006).

The images were difficult to interpret because the motions of the vertebrae proved to be fairly complex, but the muscle recordings from the *transversospinalis* muscles (the hundreds of short muscles close to the vertebrae on their dorsal surface) proved to be quite suprising. Rather than alternate right and left as I had expected, I found that both sides fired at the same time. Furthermore, there were two activations in each gait cycle. The timing revealed that the back muscles were activated as the diagonal pair of limbs were lifted off the ground. You can imagine that when the right leg is lifted, the pelvis wants to fall downward on that now unsupported side. This would tend to twist the spine. At the same time, the left forelimb is lifted, and the chest and shoulder want to fall downward on the now unsupported forward left side. This also tends to twist the spine, although it twists it in the opposite direction. At the same time, however, the leg and forelimb that are on the ground are providing unbalanced forward thrust. This tends to bend the spine as it is being twisted.

The Role of Spinal Musculature in Mammalian Locomotion

Apparently, when the back muscles fire, they provide a linear compression of the spine, jamming the joints together into a closely packed configuration. This stiffens the spine and enables it to resist twisting and bending. Because the spine is flexible, however, an animal can use the spine to adjust the relative positions of the shoulder and pelvic girdles. Even when it is bent out of linear alignment, the spine should still be capable of stiffening. To accomplish this, most of the repositioning takes place across the diaphragmatic joint in mammals (the joint between the thoracic cage and the lumbar spine).

The thoracic spine is generally kept stiff at all times. The lumbar spine can be very flexible at rest; however, in most mammals, when it is loaded with the weight of the body, the lumbar spine stiffens automatically by means of a passive or non-muscular interaction. This is accomplished by several different mechanisms, including complex arrays of bony struts and elastic ligaments, complex bone structures, and locking joints.

Rapid Alternating Transition From Flexible to Stiff

The typical motion of the mammalian spine during walking involves a simultaneous lateral bend and vertical bend. We can start the cycle with the center of the spine bent upwards and to the left. It descends into a flat linear position and then bends to the right as it bends upwards again. It is somewhat counterintuitive to appreciate that the main muscle firing of the epaxial muscles of the spine actually takes place when the spine is maximally bent and twisted into an upward, laterally displaced arch. At that moment in the cycle, the muscles on the right and left side fire in unison. The result of the muscle firing is to pull the dorsal or arch structures of the spine more tightly together, which momentarily stiffens the spine against any further rotative torques or bending forces.

Typically we think of muscles contracting in order to move or bend two bones across a joint, but in this setting, the muscle firing is carried out to stiffen the structural element by loading it longitudinally across each of the individual interspinal joints. A mass contraction of a muscle attached only at the cervical and sacral ends of the spine would bend it. However, the contraction of the small short transversospinalis muscles across each spinal joint, particularly when activated bilaterally, actually resists bending.

Breathing While Running

Looking at the oxygen flux graph (Figure 8-2E), one key conclusion seems to jump right off the page: If an animal is going to run quickly, it must be able to ramp up its breathing as inexorably as it ramps up its speed. An animal whose breathing gets less and less efficient the faster it runs will not perform well, according to this graph. In fact, that is exactly the basis for the two different lines—one for cold-blooded lizards, whose oxygen flux is strictly limited, and another for mammals and birds, both of which can demonstrate an enormous capability to increase their oxygen flow as their running or flying speed increases.

The mechanics behind this ability are a matter of two major components. The first is an optimized chest wall design. The second is a means of separating running from breathing. In some of the fastest animals, the mechanics of running may actually be used to help pump air in and out of the lungs rather than inhibiting the process. To achieve and support their high maximum oxygen flow rate, warm-blooded amniotes always have a very large surface area for the lungs, and they have efficient means of moving air while running or fighting. For the most part, this means having a relatively rigid thorax or rib cage with considerable ability to provide optimal airflow during maximal activity. The mechanics of breathing while running may have been the real battleground between mammals and the dinosaurs.

Coherent Rib Mechanics in the Chest Wall

The single-headed ribs of the lizards and snakes don't play a helpful role in keeping the ribs aligned with each other during breathing. A single-headed rib can move in any direction around its attachment point. When a lizard is running, its spine undulates and the rib movements become chaotic and ineffective. Similarly, turtles have a rigid body

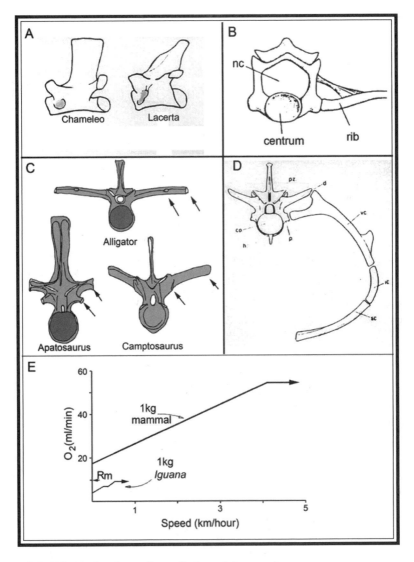

Figure 8-2: Rib mechanics and ventilation while running.

A) B) Lizards often have single-headed ribs that are pulled dorsally for breathing, a pattern severely disrupted by body undulation in running.

C) D) Archosaurs have a more stable front-to-back rib traverse with two heads. In the upper thorax, the axis of rotation is at 45 degrees, but in the lower thorax the ribs rotate on a low axis (as in many mammals), and their coordinated movement is very resistant to body motion.

E) Oxygen consumption goes up with speed, but a mammal can run much faster than a lizard because it can move far more air through its lungs.

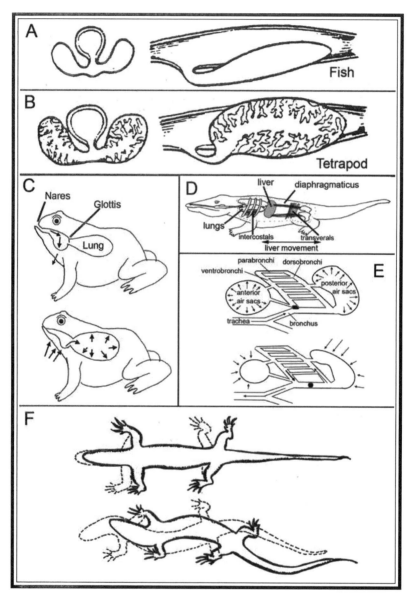

Figure 8-3: Evolution of the lung.
A) The air sac used for ballast is exploited for land breathing in *Polypterus*.
B) More lung surface area from internal branching in early land vertebrates.
C) Amphibians use a throat pouch to push air into the lung.
D) Archosaurs (alligators) use a liver-based suction pump to add to rib-based suction.
E) Birds have continuous forward air flow through their parabronchii.
F) Undulation of the body during running interferes with coordinated rib breathing.

cavity, but a relatively inefficient means of expanding the cavity to achieve ventilation. This just isn't part of their design.

By contrast, dinosaurs and mammals typically have a two-headed rib. The two heads provide a fixed axis for motion. This means that each rib moves along a very specific course relative to the vertebrae to which it is attached (see Figure 8-2D). If all the vertebrae of the rib cage can be kept aligned with each other, then all the ribs will swing together in effective unison. Indeed, the hallmark of the warm-blooded amniotes appears to be a spine that can be kept rigid while running or flying.

In addition, all of the warm-blooded amniotes appear to have some sort of secondary mechanism for expanding the chest wall. In mammals it is the diaphragm. In alligators, which are relatives of the dinosaurs, there is an interesting system of muscles that can actually pull the liver toward the pelvis in order to help expand the lungs (see Figure 8-3D). Although birds have a supplementary ventilation system involving the pelvis, their more complex airflow dictates a very specific major role for sternal and costal ventilation. It is not yet clear exactly how the dinosaurs supplemented their rib cage expansions.

A Role for the Fine Structure of the Lung

In mammals, the air is sucked in through the trachea and passed through extensively branching bronchial tubes until it finally fills huge numbers of tiny elastic sacs called *alveoli*. It is estimated that there are more than 500 million alveoli in the human lung, providing a surface area the size of about 100 square meters. When the chest is expanded, a negative pressure is produced. The alveoli expand and air is drawn into them. On the outer surfaces of the alveoli are tiny capillary blood vessels that exchange oxygen and carbon dioxide with the air inside the alveolar sacs before the muscles relax, thus allowing the alveoli to shrink and expel their used air during exhalation.

Monotremes (egg-laying mammals) demonstrate the virtually complete development of the typical lung structure of the modern mammals. The lung has complex, branching bronchi and alveoli that are very similar to those in the other mammals. In addition, the monotremes have a fairly typical mammalian diaphragm. This results in a complete separation between the chest (pleural space around the lungs) and the abdomen (peritoneal space around the abdominal organs). However, in the other amniotes (lizards, snakes, alligators, birds, and probably dinosaurs), the lung structure is totally different. There are no elaborately branched, elastic alveoli. Instead, the lung is essentially a hollow sac with more or less rigid *septations* (dividers) extending into

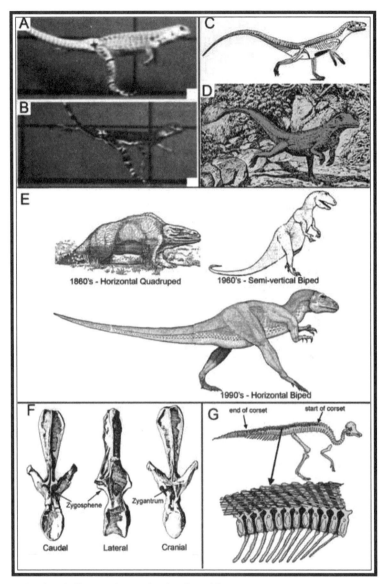

Figure 8-4: Bipedalism and running in diapsids.
A) B) To move more air, some lizards run bipedally so they can suppress undulation in their anterior body. C) D) Early dinosaurs were small and ran bipedally, probably for the same reason. E) Our understanding of dinosaur posture and locomotion has advanced steadily. From McLoughlin 1979, pgs. 7, 50, 51. Copyright 1979 by John C. McLoughlin, reprinted with permission from Viking, Penguin (U.S.A), Inc. F) When a dinosaur holds its body horizontal, the vertebrae interlock to provide a rigid beam so the ribs will swing in a coherent fashion. G) Some types of dinosaurs had ossified (bony) ligaments to hold the spine stiff.

the interior to increase surface area. This type of lung does not achieve the same efficiency of contact between inhaled air and gas exchange surface that is provided by the mammalian lung structure.

In contrast, the bird lung is extremely efficient and far surpasses the capability of the mammalian lung. The avian lung is divided into a vast number of tiny rigid channels. Air is pushed continuously through these rigid channels, producing very efficient extraction of oxygen on a continuous basis (see Figure 8-2E). However, for any amniote that does not have the flow-through mechanism of the birds or the alveolar system of the mammals, other means of efficient ventilation and increased lung surface area are required if they are going to achieve the high oxygen flows required for prolonged high speed running.

Lizards that run on two feet

One of the interesting zoologic tidbits here is provided by the various species of *cursorial* or running lizards. Although they are typical four-legged lizards most of the time, they can actually rear up onto their hind limbs and run bipedally (see Figure 8-4). Apparently, by disengaging the arms from locomotion while running, they can suppress undulation of the spine, hold their upper torso rigid, and achieve an improved mechanism for lung ventilation during running.

As we will see later, this solution plays an important role for dinosaurs. For mammals, however, a different anatomic solution emerged that has allowed for a rigid chest cavity during running. This is the separation of the spine of the body into the thoracic and lumbar regions.

The Rigid Rib Cage of the Mammals

When a mammal runs, most of the rib cage is held in a rigid straight line. Each rib has two heads—one above and one below for a two-point, hinged contact with the spine. In fact, in the majority of mammals, the lower head of the rib is placed between two vertebrae so that it contacts both of them. This helps maintain the rigidity of the thoracic spine. Finally, in most mammals, the ribs don't occur lower than the bottom of the lungs. The lower thoracic and lumbar regions can be very flexible, and can go through tremendous motions without affecting the upper ribs during running. Because of the high metabolic rate of a resting mammal, there is no possibility of running without breathing. Mammals are designed to dramatically increase the rate of breathing and increase the volume of air moved with each breath, no matter how fast they run. The two-headed rib and the rigid thoracic spine are the key to this.

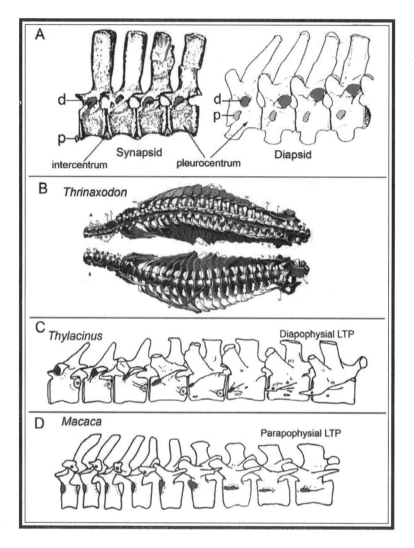

Figure 8-5: Rib articulations in the amniotes.

A) In the synapsid ancestors of the mammals, the ventral rib head articulated on a small intercentrum between the adjacent main vertebral bodies; in diapsids, this moved onto the body of the main pleurocentrum.

B) Ribs occurred on all cervical thoracic and lumbar vertebrae in *Thrinaxodon*, a cynodont synapsid close to the origin point of the mammals.

C) In a marsupial, the upper rib articulation (diarthrum) descends to help form the lumbar transverse process (LTP).

D) In most primates, the posterior half of the lower rib articulation forms the base of the LTP.

Separating the rib cage from the head and the pelvis

To assure that the ribs can move consistently in unison no matter what the rest of the body is doing, it is necessary that the rib cage be definitively separated from the head and the pelvis. In fish, there is no neck per se, as the ribs stop only at the skull. Amniotes, however, do have necks, and these are most well-developed in those with the highest metabolic rates (that is, dinosaurs, birds, and mammals). From the point of view of high-capacity breathing, this will be important if the head is used to shake, bite, and/or subdue prey. This also insures that if the running mammal has to look around as it runs, there will be no cost in breathing efficiency. Neither the cycling motions of the rib cage nor the oscillations of the running body will force the head to move excessively in a way that could impair vision or hearing.

Except in the great sea mammals such as the whale, a neck is a standard feature of a mammal. The emergence of a distinct neck among the ancestors of the mammals can be tracked by the history of ribs on the first few vertebrae below the skull. One striking feature of the ancient, mammal-like amniote *Thrinaxodon*, when compared with modern mammals, is that there are similar ribs on all the vertebrae between the neck and the sacrum (see Figure 8-5B). Most modern mammals differ from *Thrinaxodon* in that they have lost the ribs in the lumbar and cervical (neck) regions.

It is therefore fascinating to see that the monotremes (egg-laying mammals) also have ribs in the cervical and lumbar region, just as in *Thrinaxodon*. Aside from monotremes, the remaining modern mammals are the metatherians (pouched marsupials) and the eutherians (true placental mammals, including humans). All of these have ribless lumbar and cervical regions. The importance of this division of the anatomy is that it breaks up the interaction between the rib cage and the pelvis, and completely frees the head from the motions of the chest wall.

Isolation of the ribs from movements of the thoracic spine

The mammalian feature of ribless cervical and lumbar vertebrae is readily observed. However, a more subtle aspect of the isolation of rib movement in mammals escaped notice until I made a discovery in the course of my own Ph.D. thesis research. I found that all of the existing mammals also have a modification of the vertebrae that essentially lifts the spinal musculature off of the ribs. This structural feature of the vertebra—which I called the *laminapophysis*—is an extra process or projection that passes above the upper attachment of the rib (see Figures 8-5 and 8-6). At the very least, this means that when the muscles along the spine are activated for locomotion, they do not need to pull directly on the rib heads. *Thrinaxodon* does not have this modification, but the monotremes do.

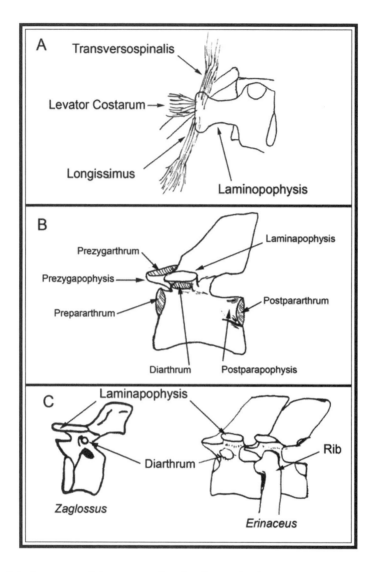

Figure 8-6: Structure of the mammalian laminapophysis.
A) Muscle attachments on the laminapophysis.
B) Naming of vertebral structures in mammals.
C) Separate diarthrum (dorsal rib articulation) and laminapophysis in a
monotreme and a hedgehog.

An Overview of the History of Mammalian Breathing

Based on this analysis, and upon comparison of the distribution of different features among the various types of animals, we can assemble a history of respiratory evolution for mammals. The oldest distinctive step appears to be the placement of the rib head between vertebral bodies to suppress lateral undulation. Next came the development of the diaphragm, and then the alveolar lung with its multiple branches, which is strictly dependent upon rigidly uniform expansion of the chest wall. The fourth step was the modification of the thoracic vertebrae to lift the locomotor musculature away from the swinging ribs. The loss of the cervical and lumbar ribs was the last step.

The point in time when all of these steps were completed is quite interesting. It seems that most of the changes took place by about 220 million years ago—quite early in the "Age of Dinosaurs." The final change—the replacement of lumbar ribs by the lumbar transverse process—seems to have occurred in the late Cretaceous period, around 80 to 90 million years ago.

The Abdominal Pump

There are two other interesting question regarding mammalian breathing that remain unresolved at this time—namely, the degree to which mammals actually use running movements to aid in ventilation, and the degree to which the rib cage serves locomotion as opposed to ventilation. The first idea is that, as the animal lands on its two front feet, the pelvic contents are driven forward, helping to push up the diaphragm. Then, as the hind limbs push off, the abdominal contents lag behind, pulling the diaphragm downwards. In fact, in many gaits, mammals do tend to synchronize their breaths with their running steps. In humans, the upper limbs do not play a large role in locomotion, so we see the opposite extreme. The two heads of the rib are aligned along an axis so that "bucket handle" rotations of the ribs support an extensive expansion of the rib cage when the intercostal muscles are contracted. It seems unlikely that there is any significant contribution of an abdominal pump for breathing when humans run.

There is also evidence that mammals use their diaphragm to enforce isolation of respiration from locomotion. Mammals can breathe efficiently when their breaths and running strides are either unsynchronized or operated on a 2:1 basis, rather than a strictly synchronized 1:1 basis (Bramble 1983, 1993). The intercostal muscles also have the task of rigidifying the rib cage so that the diaphragm can work efficiently.

From the Ribs of Bats

At yet another extreme are bats. In these mammals, the rib heads are either designed to inhibit all motions of the thoracic spine and ribs, or are actually fused to the thoracic vertebrae. Similar to birds, the flight of bats allows no room for stopping wing activity if the animal becomes tired. Further, the powerful flapping activity of the wings is a challenge to the immobility of the thoracic rib cage. Apparently for these reasons, bats typically rely on an extremely rigid rib cage.

How Birds Breathe in Flight

In mammals, the expansion of the chest draws air into the alveolar sacs, and these expand as they fill with air. Gas exchange takes place during inspiration. Then the mammal breathes out, blowing away the used air during an expiratory phase. Relatively little useful gas exchange from the blood takes place during expiration. This system works well for mammals, but it means that for at least half of the time, the respiratory capabilities of the lungs are not used. The design of respiration in birds, by contrast, results in a level of respiratory performance that goes well beyond what is seen in any mammal. Their need for very high oxygen intake has been met by a "flow-through" system that allows for continuous gas exchange. Birds do not have alveolar sacs; instead, they have respiratory air passages (*parabronchii*) that rely on a set of billows-like air chambers that produce a continuous forward flow of air through the respiratory exchange air spaces (see Figure 8-3E). In consequence, birds have the advantage of continuous gas exchange, surrendering nothing to the respiratory cycle.

This system requires an absolutely rigid thoracic space unaffected by the flapping of the wings during flight. Most birds have completely fused thoracic spines with no possibility of movement among the vertebrae. This is particularly important because the sternum serves as an attachment point for the flight muscles.

The Running Secrets of Warm-Blooded Dinosaurs

Undeniably riveting to the human imagination is the drama of a giant *Tyrannosaurus rex* shaking the earth with its enormous strides as it charges across the ancient plain of the Cretaceous era to tackle a great armored *Triceratops* fleeing desperately before it. A ferocious battle ensues, after which a horde of smaller carnivores pester the *Tyrannosaurus* as they try to steal its prey. If dinosaurs were cold-blooded, they may well have lived like lizards, spending most of their time at rest with occasional short bursts of activity. However, if they were warm-blooded like mammals, then the scene described here would bear an uncanny resemblance to what takes place on the modern African savannah as the lion chases the wildebeest. Hence, it is both our fascination with the world of the dinosaurs, as well as the usual imperatives of biological science that have motivated the exploration of this question: Were the dinosaurs warm-blooded like the mammals and the birds, or cold-blooded like the lizards and crocodiles? I believe that the answer to this question can help explain why many dinosaurs ran bipedally, rather than on all fours the way most other land vertebrates do.

The evidence that dinosaurs were warm-blooded comes from what we know of their anatomy, their body mechanics, and their surrounding ecology. There is no longer much controversy about the fact that many dinosaurs had a greater capability to increase oxygen flux than their cold-blooded relatives among the lizards. However, there are still many questions about the respiratory mechanics of dinosaurs. Fossil evidence has recently been unearthed that supports the existence of bird-like "flow-through" in the bones of some dinosaurs, which suggests parabronchial respiration. Some of the features of the respiration system of birds have now been clearly identified in the bones of early theropod dinosaurs—the ancestors of *Tyrannosaurus rex*.

The spine has not been fully investigated as a possible clue to dinosaur metabolic rates. Some recent authors have asserted that the airspaces in the vertebral bodies of some dinosaurs are equivalent to the airspaces in the vertebrae of some birds. This has been taken as evidence for "flow-through" parabronchial respiration in dinosaurs and has been the cornerstone of some arguments in favor of the warm-blooded dinosaur theory. However, there may be quite a bit more to be learned about this debate by further consideration of the spine.

Birds have a very rigid thoracic cage, often with complete natural fusion of all the vertebrae. This helps support the airflows required for their continuous high metabolic rates when flying over great distances. So what about dinosaurs? Are there other clues in their spines that will place them definitively with either the hot-blooded or the cold-blooded amniotes? Dinosaurs are divided into two main categories, based on their pelvic anatomy: the *saurischia* (lizard-like), and the *ornithischia* (bird-like). The saurischians include the *sauropodomorphs* (huge, quadrupedal plant-eaters), and the *theropods* (bipedal running carnivores such as *Tyrannosaurus rex*). Despite contradictory evidence from the pelvic bones, birds are now thought to be descendants of theropod dinosaurs, and it is this group in which there is evidence of "flow-through" parabronchial breathing.

There is also evidence from the spine that focuses the more general question of oxygen flow during running. Recall that a number of lizard species run bipedally. I have argued that this allows them to improve their oxygen flux capability during running by allowing them to breathe more effectively during periods of extreme physical exertion. The evidence for rigidification of the spine in fast-running dinosaurs is also compelling. In saurischians such as *Tyrannosaurus*, the vertebrae have an elaborate system of extra contact surfaces called *zygosphenes* and *zyganthra* (see Figure 8-4F). The result of these surfaces is that when the vertebrae are pulled into extension, they lock into a rigid beam. The implication of this is shown by the three different interpretations of the skeleton of *Tyrannosaurus* shown in Figure 8-4E. When these theropods extended their spines, they rigidified the entire body cavity for running. However, the spine could also be instantly converted into a very flexible system when it came time to fight and kill their prey.

A very different system of spinal rigidification is seen in the ornithischian dinosaurs. This group includes the *Ornithopoda*, unarmored bipedal herbivores. In many of these animals, there was an extensive corset of ossified spinal ligaments that held the spine rigid. It is possible that this served respiratory mechanics when they ran to escape carnivores, and accommodated the implications of a large, herbivorous stomach complex. The ossified portion of the corset typically begins in the mid-body region and continues past the sacrum and partway into the tail. It clearly provides a solid beam for the lower spine over the pelvis. This would certainly have uncoupled the chest region from the abdomen and pelvis during running. If these animals also had a pelvic piston system for ventilation, as there is in alligators, then rigidification of the lower spine would have served to optimize the efficiency of the ventilatory piston system.

Transformation of Repeating Stuctures and High Order Competition

Independently, in both mammals and dinosaurs, the key to existence and survival in a 160-million-year battle for supremacy appears to be due to a founding event that transformed their anatomy. Very possibly it was the *Pax* genes that control the dorsal to ventral (front to back) organization of the repeating elements (the vertebrae/somites) that were altered to move the ribs dorsally relative to the vertebrae. The consequence of this change in body plan—namely, the ability to progressively increase breathing during running—is the real basis for existence in both groups. In the Late Cretaceous, an additional change in the lumbar region in what became the therian group of mammals may have further increased the running efficiency of mammals, and finally helped tip the balance in the long stuggle so that the mammals won out in the end.

Foundation and Fate: Anomaly at Moroto

Introduction—A Revolution in Origins

In the opening scene of the Stanley Kubrick/Arthur C. Clarke masterpiece, *2001: A Space Odyssey*, a totally incongruous black obelisk appears in the ground of an ancient African plain millions of years in the past. Its sudden arrival changes everything. Before it appears, the apes living nearby are mere animals. Afterward, however, although they have changed little in outward appearance, the apes that find the obelisk are transformed in a way that seems to make them human. In this artistic context, the obelisk symbolizes our desire for a discrete and definitive dividing line between what is human and what is animal.

The discovery of a single, ancient lumbar vertebra near the slopes of the Moroto volcano in East Africa bears many of the hallmarks of that obelisk. It is almost completely incongruous. Nearby, paleontologists have found the remains of numerous species of the early apes that differed little in appearance from monkeys. That vertebra, however, bore an amazingly similar appearance to a lumbar vertebra of a human. That vertebra changes everything.

In fact, the Moroto vertebra is a surprise only because of our existing theories and models of human evolution. As we know, the vertebra itself cannot be out of place; therefore, it represents unalterable truth. What must be incorrect is our expectation of what the structure of the spine of an ancestral ape of the African Miocene era should be.

A New Body Plan for a New Type of Primate

It is now clear that 21 million years ago, in what is now equatorial Africa, a mutation occurred that created a new type of body plan: the hominoid body plan. Elsewhere in this book we have seen how changes in the *Hox*, *Pax*, and other major morphogenetic gene groups create new types of animals by creating new body plans. Apparently, this is exactly what occurred at a critical point in our own ancestry, when the lineage of the hominoid apes and humans came to be different in design from the lineage of the monkeys.

Virtually all existing models of the events that led to hominoid and human anatomy have followed the classical Darwinian pattern of gradual change. Painstaking attention has been paid to explaining ecological changes that would drive the gradual modification of a monkey-like ancestor toward the greatly modified human form. However, in the updated model of evolutionary change explained in earlier sections of this book, we expect to see occasional major changes introduced by mutational drive that reset the situation in ways that do not result directly from environmental pressures. These result from mutations, or from selection at the level of the High Order Module rather than the organism. Darwinian evolution then steps in to gradually fine tune, optimize, and vary the new type of organism. This revised model of evolutionary change seems most applicable to the events surrounding the origin of primates of human-like form—the *hominiforms*.

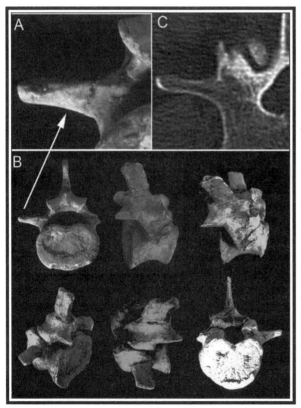

Figure 9-1: The Moroto vertebra. The lumbar transverse process (LTP) (A) of the lumbar vertebra of *Morotopithecus bishopi* is located near the facet joint on the arch of the vertebra; this is remarkably similar to the modern human vertebra (C) (CT scan). The styloid process is absent (see Fig. 9-2), and the LTP attachment reaches above the pedicle and has the unusual human homology. The pedicle is very strong, as in humans, reflecting the need for force transmission from the LTP to the vertebral body. These are key features of the anatomy of human upright posture and bipedalism.

Homeotics and the Hominoids

Although classical models of Darwinian selection can explain most of what took place in human evolution, it cannot explain the most critical initial event. When small changes in the *morphogenetic* (embryo-shaping) genes cause major, abrupt changes in numerous anatomical systems in a single generation, we need to look to the Modular Selection Theory to understand what has taken place.

In this case, there is reason to believe that the interaction of *Hox* and *Pax* genes was abruptly reconfigured to produce a new type of creature. This was an ape that was upright, broad-chested, and tailless, with an oddly caved-in lumbar spine and solid fused sternum. All of these changes occur together in the single mouse mutation called *undulated* (Gruneberg 1950, Gruneberg 1954, Filler 1986, Balling et al. 1988). We now understand that this mutation affects the *Pax* morphogenetic genes (Wallin et al. 1994, Peters et al. 1999, Mansouri et al. 2000, Kokubu et al. 2003). The modified lumbar spine in the pioneer hominiform lacked the basic anatomy of spinal support that exists in nearly every other type of mammal. In fact, it was the spine of an upright, bipedal animal. The embryological assembly of this new type of primate incorporates one of the most dramatic alterations that has occurred in the course of mammalian evolution. This is, in essence, the grand event that was the birth of the basic shape that we recognize as human.

Because we now know that a single mutation affecting the *Pax1* gene can cause all of these changes in concert, the possibility of a single *pleiotropic* event (with multiple changes from a single mutation) as a key event in our origin becomes imaginable. Indeed, as I will argue, it is not only imaginable but compelling. Once the anatomy and morphogenesis are fully understood, there are very few other viable explanations for what appears to have taken place.

In most vertebrates—fish, amphibians, reptiles, and mammals—the main separation plane between the dorsal (back) and ventral (front) of the body passes in front the spinal cord. In that strange, new primate of 21 million years ago, however, the main plane of the body abruptly swung to the other side of the spinal cord. The stunning evidence of this astonishing transformation is readily seen in one of its echoes—namely, the repositioning and reconstruction of a feature called the lumbar transverse process (LTP) (see figures 9-1 and 9-2). The remarkable impact of this transformation on spinal function is described here.

When the ancestors of the hominoids became a separate group (or *clade*), the first sign of this was a change in the shape of the molars. Outwardly, the hominoids still looked very smilar to monkeys. However, in this group of primates with five-cusped molars in a "Y-5" pattern, a second event occurred that effectively established the hominiform (human-like) body plan in an instant. The consequences of this change, and the permissive ecological situation in which it occurred, should now serve as the new focal point for understanding the origin of the upright human form. It is possible that a chromosomal speciation event, following the model described in Chapter 4, captured the transformation, and the die was cast. A reasonable argument can now be made that this one abrupt event, more than any other, is responsible for the unique current status our species has now achieved. Hands for carrying and working, long legs for standing and walking, and symbolic verbal language for communication—all of these ultimately follow from the anatomical fixation of upright posture that resulted from that single homeotic transforming mutation in a Miocene East African primate.

Figure 9-2: Critical human feature: unique lumbar transverse process and styloid loss.
A) The LTP (lumbar transverse process) in humans differs tremendously from related primates. It is dorsal to the position of the spinal canal, and thick and strong instead of flat and thin.
B) C) Styloid comparison. Lateral view of lumbar vertebrae of human, macaque monkey and *Proconsul africanus*. The human vertebra, like that of *Morotopithecus*, appears to

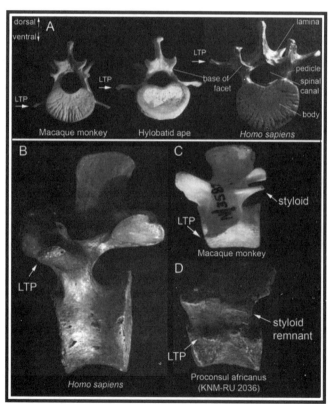

demonstrate an absence of the styloid process, and relocation of the LTP onto the arch of the vertebra at the base of the structure that carries the facet joint.
D) *Proconsul africanus* appears to have the more primitive LTP and styloid seen in monkeys and many other mammals.

The Bipedal Hominiform Hominoids

This chapter is mostly concerned with what happened between the time of the upright Moroto ape of 21 million years ago, and the time of the last common ancestor of chimpanzees and humans about 6 million years ago. The ancestral Moroto lineage was upright. The descendant *hominine* lineage—a group principally including *Australopithecus*, *Homo*, and some closely related species that have only humans as possible descendants—was also upright. Why do most paleoanthropologists (specialists in human evolution) prefer to insert a quadrupedal (walking on all fours, not upright) stage between these two?

Enthusiasts of this alleged quadrupedal ape stage seem to be enamored of the concept of the magnificent human form arising spectacularly from the crouching, animal ancestry of the lowly ape. In part, they are making the point of the split of our hominine lineage from the quadrupedal chimp and gorillas, rather than from the upright, arm-swinging gibbons whose ancestors diverged from our own at an earlier time. Most of these scientists believe they are working from evidence. However, in the absence of evidence, we can see that the scientist often works from what amounts to "scientific culture" resulting from widely accepted theories. Often, the power of adherence to old theories prevents scientists from seeing conflicting evidence. The purpose of the following chapter is to make that evidence clear and unavoidable. The theories will need to be changed.

The fact is that there is no truly compelling and definitive evidence for this interposed quadrupedal ape stage. Every bit of anatomy or fossil evidence ever proposed to support it now appears to support the uninterrupted upright ancestry instead. A group of anatomists and primatologists from the Stony Brook campus of the State University of New York—John Fleagle, Jack Stern, and Randall Sussman—have been among the strongest advocates of the view that hominine bipedalism is quite distinct from the various types of ape locomotion. However, with one flaw after another and one fracture growing into the next, a repeating drumbeat of reports and findings inconsistent with the orthodox Stony Brook view are now causing it to crumble. What is emerging instead is a picture of a persistently upright bipedal line of ancestors, extending in unbroken series from its origin in an ancient homeotic mutation event on up to its most recent incarnation as our modern human form.

In the past, the existence of upright bipedalism has been a defining element of what has recently been called the hominin tribe (Begun 2004). Indeed, the onset of upright bipedalism as the primary mode of posture and locomotion has been accepted as equivalent with onset of humanity, when it occurred and became fixed in our direct lineage. The category of "human" has included the various species of the genus *Australopithecus* and the genus

Homo, both of which are accepted as being something other than apes. Recently, the longstanding agreement in the scientific community on how we define the terms "ape" and "human" has been in a state of flux (Tuttle 2006a). For the purposes of this book, I will rely on the new term *hominiform* to describe the various hominoids—ape or human—that share the common upright bipedal body plan that appears to have emerged 21 million years ago.

The term *hominiform hominoids* also has the attractive feature of defining a proper monophyletic group from a cladistic point of view (see Chapter 4). It excludes the more monkey-like hominoids such as the *Proconsul* and *Limnopithecus* species that appear to have existed at about the same time and are defined as hominoids largely because of their teeth. However, it includes all of the living apes and humans as well as their post-*Morotopithecus* ancestors. In the Early Miocene of 20 million years ago, the hominoid lineage had only recently separated from the lineage of the Old World monkeys (*cercopithecoids*). There may have been hominoids at this time that did not have a hominiform ancestor; however, something happened that led a hominiform ape to leave its fossil vertebra on the slope of the Moroto volcano 21 million years ago. The implications of the origin of that ape demand a major change in the current theories of human evolution.

Discovery of the Moroto Vertebra

In 1958, William Bishop, a geologist from Glasgow, was hired by the Uganda Geologic Survey to investigate the source of some fossils recently discovered in a volcanic region along the Uganda-Kenya border, about 100 miles west of Lake Rudolph and 150 miles north of Lake Victoria. Bishop, who became the curator of the Uganda Museum in Kampala a few years later, made a series of visits to the region and eventually identified more than a dozen locations that were studded with numerous mammalian fossils from the Miocene era (up to 22 million years ago). The last site he found in the area in the summer of 1961 was his Moroto II location.

These were fantastic fossil sites. Cataclysmic geologic events that formed the Great Rift Valley of Africa had pushed sedimentary rocks to the surface that had been formed in the forest at water's edge during the Miocene era 20 million years before. Repeated volcanic eruptions had created a series of layers that could be accurately dated. Even more spectacular was the discovery of a series of ancient ape fossils that have generally been assigned to the genus *Proconsul*. A small species was called *Proconsul africanus*, and a large species was called *Proconsul major*.

Bishop, similar to many other geologists and paleontologists, was most interested in dental and cranial fossils, as they were the easiest to identify accurately. Limb bones were interesting only if they were closely associated with the fragments of skulls. He published some prominent reports on the dental and cranial fossils in the journal *Nature* between 1958 and 1964. Other fragments of bone were set aside for later study if the opportunity arose. A few years later, two young graduate students, Alan Walker and Michael Rose, arrived in East Africa from England. They were old friends from Royal Air Force service and from Cambridge University. Both planned to work on their Ph.D. theses and had lined up jobs teaching anatomy. Looking through a few shelves full of undescribed fossils that had been collected over several years at Moroto II, Walker and Rose identified four vertebrae that appeared similar in size and shape to the vertebrae of a modern chimpanzee. In fact, they were remarkably similar in shape to modern human lumbar and thoracic vertebrae. The most stunning of these was numbered UMP 67.28—a nearly complete mid-lumbar vertebra that was assigned to the species *Proconsul major*, now called *Morotopithecus* (see Figure 9-1). This fossil did not fit neatly into the scheme of human evolution as Walker and Rose understood it. It has never fit into this scheme. In fact, it is the defining proof of a revolutionary idea. It is proof that the common ancestor of the humans and the apes probably already possessed the key anatomical modification that underlies our upright stance, our bipedal locomotion, and our complex use of our arms and hands.

Paleontologists and anthropologists have had a problem with this fossil, in part because it can't really be understood within the paradigm of gradual adaptation and change. We can understand that the lumbar vertebrae must have been modified to accommodate the upright bipedalism of modern humans. However, if human bipedalism arose gradually over millions of years, how do we explain a lumbar vertebra that was essentially transformed into modern human form at the very outset of the evolution of the ancestors of the apes? The discovery was not considered sensational at the time because its full significance was not understood. Nonetheless, the two young scientists carefully cleaned the specimen described it in fine detail, and published its description prominently in *Nature* (Walker and Rose 1968), the leading scientific journal of our time. The species has more recently been renamed *Morotopithecus bishopi* (Gebo et al. 1997), after the Moroto volcano near the site of its discovery, and after William Bishop, who made the initial find.

It has been pointed out that the large apes at the Moroto site appear to belong in at least two entirely separate genera: a large version of *Proconsul africanus* to be called *Ugandapithecus major*, whose lumbar vertebra demonstrates the transformation; and a separate species called *Afropithecus turkanensis* (Gommery 2003). When the species name of *Morotopithecus* was proposed by Gebo and his colleagues in 1997, it was certainly intended to reflect the unique status conferred by the remarkable UMP 67.28 vertebra. The implication is that this new form emerged from the proconsulid family group to establish a new genus among the hominoids—basically, an entirely new kind of animal. Based on a new understanding of the anatomy of that vertebra, it is safe to say that this was a primate that could not walk comfortably in a horizontal posture. Laura MacLatchy, a University of Michigan paleoanthropologist who has helped discover and describe hundreds of new fossils from Moroto in the 1990s, has detailed the differences between the various large hominoids found in the region: *Morotopithecus*, *Afropithecus*, *Proconsul major*, and the proposed *Ugandapithecus*. Based on her (2004) analysis (see also Young and MacLatchy 2004), I will continue to refer to the species in which the vertebra occurred as *Morotopithecus* in this book.

While Walker and Rose were working on the publication, Bishop brought the vertebra to Cambridge and showed it to David Pilbeam, who was just then completing his Ph.D. thesis (Pilbeam 1969). Pilbeam traveled to the fossil site at Moroto in 1966 to better understand the find. It was 15 years later in 1981, while I was teaching a course in comparative primate anatomy at Harvard, that Pilbeam, newly recruited away from Yale, called me into his office and somewhat dramatically handed me a cast of the fossil lumbar vertebra and asked me to try to figure out what story it told. The 1968 paper by Walker and Rose was just one out of hundreds of articles I had reviewed in my course preparations. It was a prominent publication, but it was completed before some of the truly amazing implications of the find were fully known.

The research I did on that vertebra, and the exploration of its context among the vertebrae of other mammalian and land animal species, was the basis of my Harvard University Ph.D. thesis, which was completed in 1986 (Filler 1986). News of the wider implications I had identified spread quickly, and, when Alan Walker began work on a paper about newly discovered *Proconsul nyanzae* in 1987 (Ward et al 1993), he wrote to me asking for a copy of my thesis. Time has since borne out much of what I predicted and described in 1986, and this book is a result of continuing interest in those findings.

The History of Hominoid Locomotion

The hominoid species include the apes, our proto-human ancestors, and ourselves. No living hominoid species moves in the standard quadrupedal fashion of other mammals. Each of the hominoids is unique in its means of travel, and each carries several major transformations of the vertebral body plan of other primates.

The Origins of the Hominoids

Evolution of the primates has passed through a series of grand cycles across time. The dramatic geological diversity of Wyoming mimics the topography of the Great Rift Valley of Africa; however, its window into the past is opened to a much earlier time, including the end of the age of dinosaurs. Although there is some controversy and uncertainty about exact definitions and relationships of the earliest animals considered to be primates, it seems that these initial possible ancestors of the primate lineage, the *plesiadapiformes*, were once highly dominant among mammals. There were large numbers of species and, based on the composition of fossil finds, apparently many individuals relative to other mammals. They may even have played a role in the final demise of the smaller dinosaurs.

In North America, these plesiadapid primates of the Paleocene (65 to 56 million years ago) and their apparent successors, the adapid and omomyid primates of the Eocene (56 to 36 million years ago), may well have been outclassed and driven extinct by the advancing capabilities of rodent competitors (Silcox et al. 2005, Smith et al. 2006). By 35 million years ago, they disappear altogether from the fossil record, and no primates are found in North or South America until 28 million years ago, when the ancestor of the New World monkeys (ceboids) arrived in South America (Fleagle 1999)—apparently by rafting from Africa. Although there are many species, it is possible to describe the anatomy of their spines and limbs as typical of "generalized" mammals—quadrupedal, able to walk on the ground, dig, and climb—but without remarkable specialization for any given niche. This has been typical of primates across much of the time of their existence.

The surviving species of primates in Europe, Asia, and Africa were never plentiful until the great radiation of hominoid species in Africa during the Miocene. In many locations, primates seemed to develop specialized anatomy for eating fruit in trees. However, they appear to have gone extinct in nearly every region where they have had to compete with bats. Coincidence or not, the surviving primates of more primitive aspect, such as the lemurs and lorises, have only flourished in parts of the world that are not inhabited by bats.

Beginning in the Early Miocene era of 20 million years ago, the hominoids appeared to become a successful primate group, and numerous species begin to appear. However, after 15 million years of success for the hominoids, it was the monkeys of Asia, Europe, and Africa (cercopithecoids) that experienced a dramatic increase in species diversity and overall numbers. One hominoid species after another was apparently driven into extinction, with only a few surviving: the large, ground-dwelling chimps and gorillas of Africa, the small, arm-swinging hylobatids, and the large, quadrumanual (four-handed) orangutans of Southeast Asia. Aside from these, there was a steady flow of new species and steady increase in volume of individuals in the ground-based, upright species. These were the hominines that are our most direct ancestors, including *Australopithecus*, *Paranthropus*, and finally, the various species of the genus *Homo*.

Apes of the Miocene Era

There is an ape vertebra fossil that is somewhat similar in design to the hylobatid vertebra. It was discovered on Rusinga Island in Kenya (see Figure 9-2D), and is attributed to a species called *Proconsul africanus* (other genus names for this specimen have included *Dryopithecus*, *Kenyapithecus*, and most recently *Equatorius*). If the paleontologists had been correct in placing *Proconsul africanus* in the same genus as *Proconsul major* from Moroto (now *Morotopithecus*), then these two species of this genus would have stood on either side of a great genetic divide within the hominoids. The first has hylobatid-type vertebrae, and the second has the vertebral design of a human.

Later, in the mid-1980s, the nearly complete lumbar spine of another Proconsul species was discovered at the Mfangano Island at Lake Victoria in Kenya, a short distance from Rusinga Island. Carol Ward, an anatomist from the University of Missouri, working with Alan Walker and others, has described these vertebrae in great detail. She argues that Proconsul nyanzae had six or seven lumbar vertebrae. The transverse processes, similar to those of the vertebra attributed to *Proconsul africanus*, is at the dorsal margin of the *centrum* (vertebral body), as in hylobatids. Taken together the various limb bones of the several *Proconsul* species show few signs of the specializations for upright posture, arm-swinging, or bipedalism that are seen in all the existing apes. These species have the Y-5 dental pattern, but the lumbar spine does not appear to be shortened.

Another species of Mid-Miocene ape that seems to belong in the proconsulid group is *Nacholapithecus kerioi* (Nakatsukasa et al. 2003, Ishida et al. 2004). This proconsulid is quite interesting because it seems to convincingly demonstrate loss of the tail, and has a variety of specializations that suggest it was indeed a primarily upright primate with some adaptations for suspensory behavior (for example, hanging from its arms and arm-swinging locomotion). Here again we have six lumbar vertebrae, a small styloid process, and LTPs positioned, as in hylobatids, at the dorsal margin of the vertebral body (see Figure 9-2). Based on the model I have emphasized for the basic mechanics of the primate lumbar spine, these species do show a lumbar design more suited for upright than for quadrupedal postures. I believe that this anatomical pattern and associated lifestyle constituted a "permissive" zone for the further change that occurred in *Morotopithecus bishopi*.

It may be that if the *Morotopithecus* mutation had occurred in a more strictly quadrupedal ground-walking species (with no significant role for upright postures in its lifestyle), then the individual with the mutation would have been poorly adapted for life, and the mutation would have been lost. However, when this mutation occurred in a species in which upright postures were already part of the regular repertoire, then the mutation had the opportunity to be sustained and passed along across a few generations until a chromosomal speciation event could wall it off in a new species. Indeed, this seems to be exactly what happened 21 million years ago—possibly right there on the slopes of the Moroto volcano—when *Morotopithecus bishopi* separated from the other large ape or apes whose remains are found in the same locale.

It is likely that both *Proconsul africanus* and *Morotopithecus* were ape species engaging in vertical postures. *Proconsul africanus* may have been a vertical climber, but *Morotopithecus* may very well have been an upright ground walker, a biped. Remember: If the human vertebral mutation had appeared in an individual baby monkey of the Miocene era, that individual would have had a spine utterly unsuited for its quadrupedal lifestyle. The animal would have easily been captured by a predator before it ever reproduced and passed on the change. The mutation would have been, in effect, a birth defect that sealed its fate as a non-survivor. However, if the mutation occurred in a more upright species, then the animal would have functioned perfectly well in suspensory postures such as arm-swinging. However, this individual would also have been more comfortable walking bipedally on the ground than its other family members. It would have had the *sine qua non* of natural selection; it would have had a competitive advantage.

The Hylobatian Theory of Sir Arthur Keith

The idea that the common ancestor of the humans and the apes walked upright is not new. At the turn of the previous century, one of the last great surgeon/zoologists was Sir Arthur Keith. He was a master of spinal anatomy and it was his proposal, the Hylobatian Theory, which first laid out this idea in detail. First published in 1902, his ideas are spelled out with excellent clarity, and with an impressive array and interpretation of anatomical evidence, in the Hunterian Lectures presented at the Royal College of Surgeons in 1923. This was the same forum used by Sir Richard Owen (Owen 1846) and Sir William Flower (Flower 1885b), both of whom had preceded Keith as the curator of John Hunter's remarkable collections.

The Hylobatian Theory draws its name from the small Southeast Asian apes, the gibbon and the siamang, that are in the hylobatid family. These are animals that arm-swing through the trees but walk upright on two feet when they are on the ground. Keith believed that the ancestors of humans and apes walked upright as the modern hylobatids do (see figures 9-3 and 9-4). The smallest of the hominoid apes, and the first to separate from the others in evolutionary time (15–17 million years ago), are the so-called lesser apes. The arm-swinging is actually a hominiform trick. None of the Old World monkeys can do this, and apparently the proconsulids were unable to as well. Nonetheless, a similar capability has emerged in one type of New World monkey, the atelines. Brachiation involves an upright body plan, long arms, flexible wrists, and the proper orientation of the skull upon the neck. The hylobatids are primarily tree-living creatures, and suspensory postures (hanging from their arms) clearly dominate their life activities.

Figure 9-3: Gibbon walking on vine. The hylobatid apes, such as this gibbon (*Hylobates lar*), can be extraordinarily adept bipeds, capable of types of upright locomotion that modern humans cannot match. Their most common means of locomotion is arm-swinging or brachiation. (Photos by Nina Leen, originally published in *The Primates*, by Irven DeVore and Sarel Eimerl, *Time/Life*, 1968.)

Hominoid brachiation relates in substantial part to a unique shared aspect of the hominoid arm—namely, the ability to *circumduct*, or rotate the arm extensively overhead. This is a critical motion for arm-swinging brachiation, but it does not develop in many primates that do extensive vertical climbing. For this reason, Daniel Gebo—an anthropologist from Northern Illinois University in Dekalb, who published an extensive review of primate anatomy 1996, and a multi-authored book on primate anatomy in 1993—has argued that Sir Arthur Keith was correct in looking to brachiation as the reason why we have been able to maintain our unusual arm and chest anatomy over the past 20 million years.

Perhaps because Keith was a surgeon, he differs from most anthropologists and paleontologists in his attention to the altered anatomy of internal organs, rather than attending only to the role of bones and muscles. He first observed and described brachiation in gibbons when serving as a medical officer in Southeast Asia in the early 1890s. He realized that this form of locomotion was quite distinct from how monkeys moved. Subsequently, when he dissected gibbons and monkeys as part of his research on malaria, he was astonished to find the internal organs of the gibbons arrayed in a fashion far more similar to human internal organs than those in the monkey. He believed that the hylobatian lineage had split off early in hominoid evolution, long before the separation of the humans from the great apes. However, he believed

Figure 9-4: Gibbon running styles on the ground. These apes use tripedal and quadrupedal "crutch walking" occasionally, but their fundamental running style is hauntingly similar to human running.

that the upright (*orthograde*) posture employed for brachiation was the driving evolutionary force behind a wide array of anatomical features that had originally been thought of as uniquely human.

Keith noticed that when he compared the lumbar spinal musculature of monkeys to those of gibbons of identical body weight, the monkeys had more than twice the mass of lumbar spinal musculature. He related this to his observation of the difference between the ways that monkeys and gibbons traveled between trees. The monkeys would flex their spines to bring their elbows near their knees, push off mightily with their hind legs, and hurl their thorax upward into a straight, extended position. As they traveled through the air toward their target, their bodies were nearly horizontal. Their outstretched hands would arrive first, and grasp the leaves and branches to commence the landing. By contrast, the gibbon would remain completely upright at all times. Leaps between trees sometimes used assistance of the legs, but both launch and landing were almost universally accomplished by the arms.

Although Keith also described a variety of modifications of the spinal musculature to support upright posture in humans, his most striking and least cited observations concerned the position of the diaphragm (the large muscle that forms the floor of the chest cavity and expands the lungs) and the anatomy of the heart. He observed that the diaphragm in monkeys was oriented diagonally so that it faced mostly downward as the animal stood in horizontal (*pronograde*) position. In hominoids, however, the diaphragm is parallel to the ground in the upright stance. The weight of the viscera tend to expand the chest cavity in the horizontal monkey (also the result of the active phase of muscular contraction of the diaphragm), but this effect only comes into play in hominoids when they are in a vertical posture. Further, in the great apes and in humans, Keith observed that the heart was broadly attached to the diaphragm. In monkeys, the heart swings forward to hang downward as they shift into a horizontal posture. The inferior vena cava has a mobile segment in the thorax so it can follow the movements of the heart. In hominoids, however, the inferior vena cava has no free segment in the thorax; thus its conduit flows normally into the right atrium, in a fixed relation to the similarly fixed heart.

Despite all of the accumulated evidence, Keith found few supporters for his view. By the 1940s, impressed by the similarities between humans and the African great apes (the chimpanzee and gorilla), he ceased advocating his hylobatian model at all.

The Hylobatid Lineage: Gibbons and Siamangs

Although the concept of brachiation and bipedal walking has been well-accepted for the hylobatids, Evie Vereecke, an anatomist from Belgium, and her colleagues have detailed the frequency and styles used by gibbons in tripedal and quadrupedal progression (2006). Their arms are so long relative to their body and legs that they remain fully erect when the hands touch the ground. They can walk bipedally at a rate that is equivalent to the human jog or run. Above this rate, they can use one or both of their arms as crutches. When two arms are used, they may plant the two arms and swing through between them. The tripedal gait involves using one arm to include the occasional assisted leap as part of their running style (see Figure 9-4).

When the advocates of a knuckle-walking phase see the remarkable bipedal repertoire of hylobatids, they have to insist that this spectacularly unusual behavior—so similar to our own walking and running—emerged independently of the bipedalism in humans by a remarkable coincidence. A scientist may not be willing or able to give up a lifelong career belief in a knuckle-walking phase from which human bipedalism arose *de novo*. However, it is difficult to sustain a convincing absolute position against any connection. Is it really totally impossible that there is a direct relationship between the bipedalism of humans and the bipedalism of hylobatids? The possibility must be admitted.

Jack Stern, Chairman of the Department of Anatomy at the State University of New York at Stony Brook, proposed that vertical climbing in trees led to a modified pelvis well-suited to walking (Stern 1975). However, Dan Gebo has since pointed out that vertical climbing makes up only a very small portion of the activity in any primate, and convincingly refutes Stern's idea (Gebo 1996). It seems that bipedalism in the earliest hominiform primate, *Morotopithecus*, is the source of our own modern bipedalism.

The Hylobatid Lumbar Reversion

The lumbar vertebrae of the gibbons and siamangs actually do not demonstrate the great mutation seen in the human vertebral form. They have lumbar transverse processes similar in shape to the ones found in monkeys; however, they are unusual in position. They are placed exactly at the posterior margin of the vertebral body (see Figure 9-2). The result is that they are not optimized for quadrupedal stance, and not fully optimized for ground-based bipedal walking, either. This is not too much of a problem for hylobatids because they usually travel by arm-swinging.

For many years it was assumed that the hylobatid lineage split off before the *Proconsul* species had existed. However, the evidence is now quite strong that *Morotopithecus* was carrying its vertebral transformation as long as 4 million years before the divergence of the hylobatid lineage (Raaum et al. 2005). This timing is now widely accepted among paleoanthropologists (Pilbeam 2004). It is therefore certainly possible that in the hylobatid lineage, the mutation that transformed the LTP was lost, leading to a reappearance of the ancestral, anthropoid, primate style of LTP. In this case, it could be argued that the adaptations to upright posture, including suspensory postures and brachiation (arm-swinging), actually occurred in descendants of *Morotopithecus* before the origin of the hylobatid lineage. This would explain the shared prominence of upright and suspensory postures among all the apes and hominines. Most importantly, it places the transformation of the body plan at Moroto as the key triggering event for the hominoid body plan. This human-like anatomical transformation of the lumbar spine, sacrum, pelvis, and chest was therefore the first event. A single morphogenetic mutation affecting the *Pax1* gene is now known that can produce these all four of these effects in unison, and indeed this is what might have taken place to launch the hominiform lineage and career.

The so-called hylobatian model of human evolution has often been rejected in favor of a much later, separate origin for human upright posture. However, the new evidence of divergence times gleaned from biochemistry, together with the Moroto vertebra, suggest just the opposite. The anatomical basis of human upright posture was first to emerge. It was this human-like ancestor that set the stage for both the hylobatids and the later *pongids* (the great apes, including the chimp, gorilla, and orangutan) that have diverged from the original, human-like body plan.

There are many competing theories about the details and about the main sequence of events in human evolution. At this point, however, no theory about the origins of upright posture is complete unless it can account for the evidence from the fossils of the spine. Discussions of the meaning of the spinal changes should at least incorporate our functional and anatomical understanding as of the 1980s. This one remarkable fossil vertebra from Uganda unavoidably transforms our insight into the history of human origins, and the relevant information must be faced—particularly because the results conflict with previous theories based on inferences from other areas of anatomy that are not definitive.

The Troglodytian Model and Sherwood Washburn

From the 1950s through the 80s, Sherwood Washburn—a famous anthropologist, author, and member of the National Academy of Science who worked initially at Columbia University, and was later Chairman of the Department of Anthropology at the University of Chicago and then Chairman at U.C. Berkeley—established and assiduously popularized the one model that has most strongly influenced the public as well as many scientists (Washburn 1950, 1978, 1982). This is the familiar tableau of a four-footed monkey, followed by a quadrapedal ape walking with its body carried at a 45-degree angle, rather than horizontally. Next to this we see an archaic hominid species, such as *Australopithecus* or early *Homo*, walking bipedally but still hunched over. The final image completing the tableau is the proud, tall, and fully erect form of a modern human.

Washburn described the long arms of the knuckle-walking, partially erect apes as if they were literal crutches. Their descendants of course gradually relied less and less on these crutches to achieve the paragon of free-standing, upright bipedalism. Gibbons do "crutch-walk" from time to time, but that is for full, upright bipedalism, and it is unrelated to how chimpanzees and gorillas use their arms. Sir Arthur Keith's model of an ancestor that was upright from the beginning essentially disappeared from academic or public discussion, and many professional paleontologists and anthropologists were never even aware that there was such a model with the substantial anatomical evidence to support it.

The Role of Knuckle-Walking

The first major scientific challenge to Washburn's model arose out of the work of a University of Chicago anthropologist named Russell Tuttle. Working from a rigorous anatomical perspective, and applying the most advanced technologies of the 1970s, Tuttle showed by electromyographic studies of the muscle activity of live walking apes, as well as from the detailed anatomy of their hands and arms, that orangutans did not regularly knuckle-walk, and that the knuckle-walking of chimps and gorillas might be a very specialized new development in those two species only, rather than an ancestral condition for a chimp-gorilla-human trichotomy.

From Tuttle's perspective in the 1970s, it seemed that knuckle-walking might have evolved in a common ancestor of the chimps and gorillas after that ancestor split from the human lineage. Since then, however, there has been a large amount of molecular data that makes it nearly certain that chimpanzees and humans are *sister groups*, or descendants of the two limbs of the branch point in the cladistic family tree (see figures 5-3 and 10-4), and that the ancestor of the gorilla split off as long as 1 or 2 million years before the chimp-human split (Ruvolo 1996, Raaum 2005). This has been taken by many anthropologists to prove that knuckle-walking on all fours must have been a characteristic of the last common ancestor of humans, chimps, and gorillas. What this and the following chapter point out, however, is that the actual basis of knuckle-walking is the *diagonograde* (diagonal) body posture employed by chimps and gorillas, and that this feature is based on very different anatomy in these two types of great ape. This diagonograde posture developed twice independently (once in the gorilla lineage and once in the chimp lineage), and there is no evidence of a history of this posture in the human anatomy.

Prospects for a "Humanian" Model

The evidence supports an upright, bipedal ancestor for all of the apes before the hylobatids split off; an upright, bipedal ancestor for the great apes before the orangutan lineage split off; an upright, bipedal ancestor for the African apes and humans before the gorilla lineage split off; and finally, an upright bipedal hominiform as the common ancestor of chimpanzees and humans. We have the human form of lumbar vertebra, clearly transformed for upright, bipedal posture at 21 million years ago in *Morotopithecus*. We have the hylobatid lineage, in which the modern descendants are bipedal walkers when on the ground, splitting off at 17 million years ago. We have the *Oreopithecus* lineage (which were themselves upright bipeds) splitting off at 15 million years ago (Rook et al. 1999). *Pierolapithecus* occurs at 13 million years ago, and appears to have the same upright spinal anatomy as *Morotopithecus* (Moya-Sola et al. 2004). Next, we have the lineage of the orangutans at 14 million years ago, with modern descendants who seem to walk bipedally more often than they employ any quadrupedal gaits when on the ground. The gorilla lineage splits off at 8 million years and leads to the modern quadrupeds. However, there is now intriguing evidence of a possible common ancestor of the chimpanzees and humans at 7 million years ago. This common ancestor was also an upright biped before the chimp-human split 6 million years ago (Guy et al. 2005, Wolpoff 2002).

For a variety of reasons it has always been considered that the tailless, upright, bipedal hominoid primates should be classified as human, or at least as what we now call hominines. There are many components to humanity per se; however, the essential body plan that has represented the stem form of hominoid for more than 20 million years may well have been the human body plan. It is in this sense that a new "Humanian" model can be proposed as an alternative that may replace the older Hylobatian, Troglodytian, Vertical Climbing, and Quadrumanual models.

The Transmutation of the Lumbar Transverse Process

The Role of Evidence from the Spine

For the most part, our theories about the origins of upright posture are based on the opinions of thousands of experts regarding certain areas of anatomy. The shape of the skull, the build of the leg bones, the hands, the feet, and even the teeth all provide the basis for the leading theories. Nonetheless, posture is overwhelmingly a feature of the spine, and the structural basis of human upright posture is deeply imprinted in the unique and remarkable shape of human vertebrae. Excepting spinal surgeons (who are rarely concerned with evolution), there are very few general experts in the field of vertebral anatomy. Perhaps in consequence of this fact, most anthropologists and paleontologists tend to leave out or marginalize the evidence of the spine when they tell the story of human evolution.

As he reconsidered the role of knuckle-walking in human evolution, Russell Tuttle (who had studied under Washburn) dusted off the widely abandoned ideas of Sir Arthur Keith's hylobatian model. He began to wonder if a revision of this theory could reinstate upright posture as a key early feature and permanent component of human evolution. It was Tuttle who, in 1977, first asked me to study the spinal anatomy of apes in search of evidence for this idea. And, when David Pilbeam showed me the Moroto vertebra a few years later, I already knew the literature very well and had a variety of new elements of understanding of primate spinal anatomy already discovered. For all those reasons, I recognized immediately that what Pilbeam suspected was very true. The lumbar vertebra of *Morotopithecus* was indeed extraordinary.

In my first attempt to understand the design of the fossil vertebra, I placed it between the lumbar vertebra of a human and the lumbar vertebra of a monkey and considered the differences. First and foremost I noticed something peculiar and quite striking about the bony projections, or *processes*, sticking out from the side of the vertebra—in particular, the lumbar transverse processes, or LTPs. If you look at a lumbar vertebra from above, the LTPs stick out to each side, left and right. In humans, the LTPs are strong, powerful projections, with a square or triangular cross section that is able to handle a great deal of muscular force. In the monkey, the LTP is a thin, flat sheet of bone that looks as if it would snap off with the slightest twisting stress. Even more striking is the difference between the way the LTPs attach to the vertebra in humans, as opposed to those in monkeys. In humans, the LTP attaches to a part of the vertebra near the facet joint, atop the pedicle and closer to the back, whereas in monkeys the LTP attaches to a part closer to the front of the animal (see Figure 9-2).

As with any vertebra, the biggest part of the bone is a large, round cylinder with the upper flat end facing toward the head, and the lower flat end facing toward the tip of the tail or the end of the coccyx. This cylinder is called the body of the vertebra. The elastic discs of the spine are placed between the cylindrical bodies, and together the disks and vertebral bodies bear the weight of the body. Behind the bodies and the disks, and closer to the skin of the back, is the arch of the vertebra—so called because of its shape. It forms a curve of bone around the spinal cord and nerves of the spine. The arch is attached to the body of the vertebra through its two pedicles that stand on the body. The roof of the arch is called the *lamina* because it is a flat sheet of bone. So, in passing from the skin of the back toward the skin of the belly, there is the roof of the lamina standing on the pedicles, the spinal cord and nerves running between the pedicles, and then the heavy cylinder of the vertebral body. In monkeys, the LTP attaches to the cylindrical body of the vertebra in front of spinal cord and nerves. In humans, however, the LTP attaches in a completely different location—high on the arch of the vertebra, in line with the lamina, and well behind the spinal cord and nerves.

Embryology of the LTP

As the embryology that underlies this change is explored, the matter gets even stranger. When a mammalian embryo is developing, the body or *centrum* of the vertebra forms at the same time as the lumbar transverse processes. These bony projections are essentially modified ribs. The formation timing of the ribs and the lumbar transverse processes are essentially identical in most mammals. However, the arch portion of the vertebra, the part around and behind the spinal cord, is formed later in the course of development. Thus, from an embryological assembly point of view, the arch of the vertebra is quite distinct from the vertebral body and transverse processes. In most mammals, the body and LTPs form well before the arch of the vertebra. The instructions for forming these two regions seem to come from a different "chapter" in the script that spells out the genetic instructions for formation of an embryo and fetus. In humans, however, the assembly sequence departs from the mammalian script altogether. The lumbar transverse processes form later in the process, along with the arch of the vertebra, and not early, along with the body of the vertebra as in other mammals (Filler 1986). This is a remarkable chronological discontinuity in the playing out of the embryological plan.

Imagine, if you will, the great and ancient sequence of the assembly of the organs and tissues of the embryo. Many of us have seen images or even time-lapse photography of this process. The embryo lengthens, segments appear, the head becomes more formed, limbs grow out, and the appearance of the adult form gradually becomes more and more evident. All of this proceeds relentlessly in a predictable sequence, one step after another, in the ordered pace of a metronome. This is a sequence played out billions of times every year as the creatures of the world prepare to be born, and it has played out an almost unimaginable number of times in our various ancestors for hundreds of millions of years. However, at one point in the development of the human form, the ticking of the clock is disrupted, and the sequence is changed. The moment arrives and the lumbar processes are not formed; then, at a much later point in the sequence and at a different location, the formation of this human spinal element inserts itself.

A Transformation of Serial Anatomy Under Field Homology

As my own work from the 1980s has been cited more frequently, and as the research by others interested in the primate spine has become more visible, discussions of the position of the lumbar transverse process have gradually become essential components of any discussion of hominoid evolution (Shapiro 1993, Ward et al. 1993, Sanders and Bodenbender 1994, Begun 2003, MacLatchy 2004, Gommery 2006). However, these discussions have tended to treat the LTP position issue as if the various species simply show varying positions of a single structure along a gradient of possible locations. This is very different from the discontinuous (sharply different) transformation that *Morotopithecus* seems to represent.

Transformation of the styloid process

Missing from nearly all of these discussions is a full appreciation of a critical second aspect of the LTP shift—that is, its relationship to the mystery of the missing styloid process. The styloid is present in most mammals, including monkeys and lemurs. It is also seen on the vertebral body of the hylobatids, as well as on the KNM RU-2036 vertebra attributed to *Proconsul africanus* (see Figure 9-2D). It is also the major attachment point for one of the principal spinal muscles, the longissimus. This is confusing for many anatomists and paleontologists for two reasons. Firstly, it is a structure that humans do not appear to have (see Figure 9-2B). Secondly, it is an example of a strange phenomenon called *field homology*. As the dorsal-ventral body plane swung from the front of the spinal column to a new position behind it, this represented a repositioning of a morphogenetic field.

A series of anatomists going back to William Flower in the 1880s (Flower 1885) have recognized the strange consequence. In most mammals, including the hylobatid apes, the LTP is *serially homologous* with the rib. In other words, although they are different in shape and attachments, the rib and the LTP are made out of the same original structure on the side of the forming vertebra. The rib is an incredibly ancient structure in the spine of vertebrate animals. As mammals evolved a distinct lumbar region, a shortened rib fused onto the lumbar vertebral body. Mammals also have a novel piece of anatomy that I named the *neomorphic laminapophysis* (NLM) when I first described it in my 1986 Ph.D. thesis. It arises late in vertebral embryology on the dorsal part of the vertebra or lamina. Viewed as a repeating series of structures recurring at each vertebra along the spine, it "splits" at the thoracolumbar transition point into anterior and posterior halves. In doing so, it gives rise to the styloid process in the lumbar region of most mammals. In humans and great apes, however, the LTP is made from the NLM and not from a rib. The serial homology of the LTP has changed.

This alteration—from a rib-based (*pleurapophysial*) to a styloid-based (*laminapophysial*) LTP—is a critical mark of hominiform origins, as well as a mark of a dramatic alteration in the embryological assembly process. This is why we can say that *Morotopithecus* demonstrates the modern human configuration of this critical anatomical change. *Morotopithecus* has no styloid process.

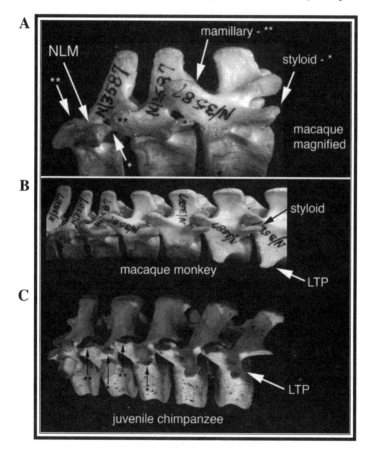

Figure 9-5: Splitting of the laminapophysis.
A) B) A key feature of mammalian anatomy is the laminapophysis (NLM) that carries the major muscle attachments of the thoracic spine. In most mammals, this structure "splits" at the thoraco-lumbar transition to yield a superiorly directed mamillary process and an inferiorly directed styloid process. The LTP of a monkey is in series with and homologous to the rib and not to any NLM-related structure.
C) In the chimpanzees and humans, the styloid is transformed into a laterally directed LTP so that the styloid appears to be missing. The splitting sequence shows that the human/chimp LTP is serially homologous to the styloid of the monkey and not to the rib. In this young juvenile chimp, the growth timing of the LTP is shown by the dark cartilage plate at its tip.

The impact of the lost styloid

When did the styloid disappear in the course of human evolution? And what was the impact on spinal function? One important feature of a styloid process is that it limits how far a creature's lumbar spine can be extended. In order for a creature to stand fully upright and arch its back into full extension, the styloids must be gone or at least greatly modified. In the hylobatid spine, the styloid is still present although greatly reduced. This allows it to continue serving as the point of attachment for the major back muscles while allowing some degree of hyperextension, particularly at the lowest vertebral levels.

In the human configuration, the styloid is gone altogether (see Figure 9-2). Remarkably enough, when the embryological sequence of human spinal development is studied in detail, it turns out that the human lumbar transverse process forms at the time that the styloid process should be forming. In fact, in the embryos of the great hominiforms (humans, chimps, gorillas, and orangutans), instead of being fashioned out of a rib, the lumbar transverse process is fashioned out of the styloid process (see Figure 9-5C).

A new anatomical role for the LTP

Not only does the human lumbar transverse process take over the embryological timing and position of the styloid, but it also takes over all the muscle attachments that go to the styloid in other mammals. This actually has the beneficial effect of moving the muscle attachments out to the side and away from the midline of the spine. This shift increases the leverage of the muscles that both cause and resist the lateral bending movements of the spine. It also brings the attachment superior to the facet joint, rather than inferior to it as in most mammals. This effectively inverts the function of the longissimus, turning it into an extensor muscle (in other primates and mammals, it usually acts as a flexor instead) by moving the line of action to a position above the axis of the joint.

Two aspects of the anatomy of the LTP that affect the forces that interact with it are the location of the tip of the LTP relative to the axes of rotation, and the location of the points of attachment to the vertebra. The location of the tip relative to an axis will determine the action of a muscle (flexion vs. extension) as it acts on the tip. The location of attachment to the vertebra will affect the translational (sliding) forces that will tend to drive the attachment point toward the axis of movement. When observed laterally, the tip of the LTP is usually at the level of the dorsal margin of the facet (see Figure 9-2). Also, the attachment of the dorsal part of the LTP is usually at the level of the ventral aspect of the facet (see figures 9-1 and 9-2). This means that when the back muscles are used in a hominiform lumbar spine to supplement extension for supporting upright posture, there is virtually no stress at all on the facet joints. This is a critical architectural feature that is identical in *Morotopithecus* and *Homo*, but is totally different from what is found in non-hominiform primates (see Figure 9-2) and the hylobatids, whose locomotion is dominated by overhead or suspensory posture and locomotion.

In most mammals and in all the monkeys, the lumbar transverse process is essentially a flat, passive element, embedded in the elastic sheet of the transverse ligament, and only employs tension in its role of tying the lumbar vertebra into the elastic support of the transverse ligament (see Figure 9-6). In humans, however, the lumbar transverse process is a strong, thick process, usually triangular or square in cross-section, but never flat. It carries some of the enormous bending force of the longissimus muscles that hold the body erect over the pelvis.

The transformation in Morotopithecus

What about the vertebra from *Morotopithecus bishopi*? No styloid process can be seen. It is fully replaced by a human-like lumbar transverse process with a stout, triangular cross section arising from the pedicle. Here again, there is a significant difference from *Proconsul africanus* because that species resembles the hylobatids in demonstrating small, separate styloid processes.

Now we need to sort through the various descendants of *Morotopithecus bishopi* to understand the eventual outcome of this critical mutation. If the common ancestor of humans, chimpanzees, and gorillas walked on two feet, how can the anatomy and posture of chimpanzees and gorillas be explained? These are the two living species most closely related to us, and yet they certainly do not move as we do. As this process of comparison and review is carried out, another very striking aspect emerges. The position of the LTP on the vertebral body varies from those on most mammalian species. In addition, among the species of living and extinct hominid apes and humans, its position on the pedicle can vary as well. However, since the time of the hylobatid divergence 17 million years ago, the LTP never crosses back onto the body from its new position on the pedicle. Once it attains its new nature as a late-developing, laminar-based structure, it never returns to an early-developing, rib-based serial homology. This is a true and definitive transmutation in the sense that one thing has, in essence, become something else. From the point of view of general genetics, this represents fixation of a trait. From the point of view of morphologic genetics, it appears to be a clue to some underlying, fundamental, and genetically traceable alteration of hominoid embryological assembly.

Functional Anatomy of the LTP

In studying the mechanics of the spine, we can see that this change in attachment point for the LTP is part of a huge change in how the lower back works in monkeys as opposed to humans (Filler 1981). Having an LTP on the cylindrical body aids the system that holds up the belly of the monkey while it stands and while it walks on all fours (see Figure 9-6). Having an LTP far back on the arch is essential to the particular anatomy in humans that helps us stand upright. It also allows for positioning of the center of gravity of the body over the appropriate kinetic balance point of the pelvis during our entire gait in bipedal walking or running. The difference in the attachment point of the LTP between monkey and human is a very visible marker of how evolution has changed the spine during the course of human evolution.

So what about lumbar function in *Morotopithecus*? In this one key feature, these creatures had assumed the form of a human, even though the rest of them—their skulls, brains, teeth, hands, and feet—differed little from the anatomy of other early apes. Perhaps this change in the spine made it much easier to stand and walk erect, and so set our ancestor on the path that leads to us. Walking more comfortably in an upright posture for longer distances enabled the use of the hands for a wider variety of activities than just locomotion; now the hands were free to carry and throw, hold on to wandering children, pull and tear, and defend and attack. All of this set the scene for more development of the brain. Basically, this change in the lumbar spine may well have set the scene for the entire human career.

This change in the spine originated as a random event. However, the persistence and spread of the trait also indicates natural selection. It is reasonable to suggest that in the parent species, some kinds of upright activities were already important. These activities may have included vertical climbing, hanging below a branch, or walking on the ground between trees. It can certainly be argued that if this change had occurred in a type of animal that had no ecological niche available to enable it to take advantage of upright posture, then it would have been an unfortunate birth defect leading to the early demise of the young individual that carried the gene. Of course, the fact that humans and *Morotopithecus* share the change in the spine helps show they are related to each other as species, but it also says something about a fundamental link in the style of life that links the ancient ancestor and the modern descendant.

Alteration of the basic mammalian design

It appears to be unlikely that any ordinary four-legged species of medium to large size could have tolerated a mutation such as this. It is a decent advantage for an upright creature that walks on two feet, but it is a severe hindrance for the more typical mammal walking on all fours. There is a large group of mammals that shares a common design for the lumbar region. This group includes the carnivores, rodents, and primates, and reflects a very ancient type of body plan dating back more than 70 million years. This body plan has been the successful basis for thousands of species across the ages.

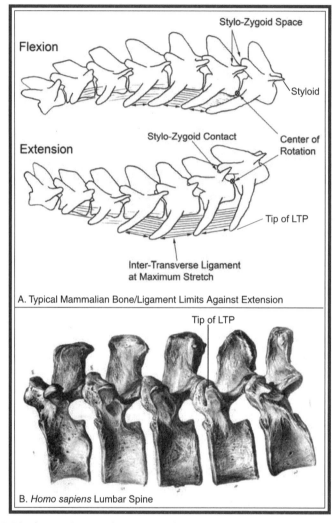

Figure 9-6: Flexion and extension mechanism in the lumbar spine.

A) When a quadrupedal animal is standing or walking quadrupedally, the weight of the body suspended between the arms and legs pulls the spine into extension. A dense, heavy, elastic "inter-transverse" ligament between the LTPs is pulled taught and resists extension, providing the usual principal support of the body when it is horizontal. Contact between the base of the styloid process and the tip of the facet joint (zygapophysis) provides a "stylo-zygoid" contact that can also resist extension of the spine. When contact occurs, the instantaneous center of rotation moves up to the point of stylo-zygoid contact.

B) In *Homo sapiens*, the intertransverse ligament system and the stylo-zygoid contact system are both lost, eliminating the standard limitation mechanisms against hyperextension. The relevant anatomy is similar in *Morotopithecus*.

Various types of Mesozoic mammals (living during the time of the dinosaurs) did not have LTPs at all. Lacking a bone and ligament support system for the lumbar region, all of the mammals for more than 100 million years seem to have been restricted to very small size—usually less than half a pound in adult body weight. The emergence of the LTP support system in the late Cretaceous period marks the dawn of the therians (marsupials and placentals), which can be hundreds of times larger than Mesozoic mammals. We know that small mammals killed and ate small dinosaurs in the Mesozoic era. Mammals apparently can run much faster than dinosaurs of similar size. Emergence of the LTP system in the therians in the late cretaceous may have made it possible to have large mammals killing and eating large dinosaurs. The fossil evidence presently available doesn't settle this question one way or another. Along with the famous meteor strike, the LTP system may have played a major role in the extinction of the dinosaurs and the emergence of the mammals as the dominant land amniotes. It is this very critical LTP system that is turned upside down in the hominoids.

This lumbar design for four-footed creatures uses elastic ligaments to bear the weight of the body, rather than relying on muscles to hold the body up during walking, running, and standing (see Figure 9-6). This is also very important for managing pregnancy, because it makes it possible for the same ligaments to support the great additional weight of the abdomen. The design is a "pivot and tension" array, similar to the design used on some types of suspension bridges. The cylindrical bodies of the vertebrae form an arch, and the "spoke and rope" design—formed by the lumbar transverse processes that point slightly downward toward the center of the arch—holds up the arch. The ropes are formed by the thick, elastic ligament that runs between the LTPs. When the center of the arch of the vertebrae is pulled downward by gravity, the spokes made by the processes have to swing away from each other. The elastic ligaments prevent this.

In the quadrupedal mammals that have transverse processes at the level of the very back or dorsal margin of the vertebral body, the elastic ligament comes under tension when the vertebral extension axis moves up toward the level of the styloid process. In other mammals, the tips of the LTP are angled toward the front of the body so that the elastic ligament is still engaged during extension of the spine, even though the base of the LTP is near the dorsal margin. In ungulates such as cows and horses, the LTP can be positioned quite far dorsally, but these animals rely on hard, bony "stops" that lock the joints of the lumbar spine and prevent extension caused by gravity (Filler 1986, 2007a). When these locks are engaged, they limit the degree to which the center of rotation can move dorsally toward the skin of the back. This is basically the same mechanism employed by orangutans and gorillas after their upright ancestor developed a new type of quadrupedal walking (see Chapter 10 and Filler 1986, 2007a).

The modified human functional design

In the human anatomical form, the lumbar transverse processes actually point towards the back (posterior), rather than toward the front (anterior), as in many mammals. They are also above the axis of rotation that exists between each pair of vertebrae. Because of these two features, there is no way to support the lumbar spine in humans in the horizontal position, other than using the abdominal muscles. Because we have no styloid processes, there is no way to limit extension available anywhere on the posterior elements (the arch, facets, and spinous processes). Horses, for example, actually impact the tips of their spinous processes to limit extension (Filler 1986, 2007a). However, the performances of modern human acrobats suggest that, despite the extraordinary extremes of extension that can be observed, our spinous processes do not lock and engage.

Finally, the human and the *Morotopithecus* lumbar vertebral designs are linked by yet another fascinating and unique aspect of anatomy that is exceptionally informative. This is the strong, box-like cross section of the LTP—often triangular or square, as opposed to the flat, blade-like shape seen in other mammals. Across hundreds of mammalian species this does not occur. This is because most of these other mammalian species have their LTP embedded in a flat membrane that plays an important role in their function. The orangutan and gorilla, for example, have modified posterior elements that block lumbar extension in certain non-upright locomotion patterns. In addition, the chimpanzee has lost the box-like cross section in favor of a reinvented flat LTP with its unique pattern of iliac suspension of the lumbar spine. The box-like cross section—when viewed in context with other mammals, and in particular when viewed in context with other hominoids—is powerful evidence that the human LTP is acted upon by powerful spinal muscles that are not "in plane" with the LTP. In humans, none of the muscles that attach to the LTP are ventral to it; they are all dorsal to it and, unlike those in chimpanzees, out of plane (that is, they are behind the LTP and pull it backward). It is this feature of the *Morotopithecus* lumbar vertebra that catches the eye of the observer of the comparative anatomy of the mammalian spine. It is the one feature that looks the most hauntingly human in its aspect.

In the upright hylobatids, whose locomotion is dominated by suspensory postures, the position of the LTP and the mechanics of the back muscles have relatively little critical impact upon their life. They typically run or walk only short distances. My own dissections of the back muscles of a variety of hylobatids reveal the remarkable underdevelopment of their back muscles compared to both what we know from human anatomy, and to what I reported in my master's thesis research from dissection of chimpanzee back muscles (Filler 1979).

What we see in humans and what we see in *Morotopithecus* is quite different. In this fundamentally hominiform architecture, there is compelling evidence of strong back muscles acting dorsally to the plane of the LTP in order to accomplish and maintain extension. There is no need for such a mechanism in suspensory upright posture. What we have here is a critical and unmistakable early signal of hominiform upright posture, with support from below rather than from above. This means an upright biped. That is the essence of the hominiform architecture of the lumbar vertebra of *Morotopithecus*.

A paleoanthropologist with an interest in the spine recently suggested that the *Morotopithecus* vertebra resembled a baboon vertebra, and gave no sign of upright posture, let alone bipedalism. However, whatever resemblance exists is limited to overall size and a partial set of shared mammalian vertebral elements. Otherwise, the vertebra is totally transformed for upright bipedal posture. Any claim that it is similar to a monkey vertebra is really indefensible, and worthy of serious consideration only for purposes of its replete repudiation.

We humans have large muscles along the anterior or front part of our spines, but they attach to the femur and cannot apply flexing force directly to the lumbar vertebrae, particularly when the hip is flexed as in a quadrupedal posture. Furthermore, although bones and ligaments never tire, back muscles do. Hence, the human spine does much better when the person is standing or walking upright than when the person is moving on all fours. This is particularly true when we try to carry anything heavy on our backs for any significant distance.

With these considerations in mind, it does seem that the change in the lumbar vertebrae had a very significant impact on the lifestyle of these human ancestors. It also seems to represent a very significant change in body plan. However, this change in vertebral structure and function is part of a whole suite of major changes. Essentially, monkeys have lumbar spines that are built just like those of many other mammals, but the spine in all the hominiforms departs from the basic mammalian plan. Among all of these hominiform hominoids, it is the human and the *Morotopithecus* lumbar vertebrae that show the clearest evidence of design for hind-limb-supported upright posture (bipedalism). The echoes of fully formed bipedalism in the existing apes reflect this shared locomotor ancestry that has been preserved in its primitive form in modern humans.

The Role of Bipedalism in the Locomotion of *Morotopithecus*

When people today look at a hylobatid swinging through the trees, their first thought is that they are seeing a monkey. They may perceive something unusual about these creatures, particularly if they happen to see one come down from the tree and run along the ground to reach the next tree (see Figure 9-4). Even more startling is a view of a gibbon running bipedally up a vine (see Figure 9-3). In this matter, we appreciate that these animals are better at some types of bipedal locomotion than modern humans.

All apes and humans can arm-swing (brachiate), which is something that most monkeys cannot do. The only exception among the monkeys are a few species of New World monkey related to the spider monkey (*Ateles*), all of which have independently evolved many of the traits seen in the apes—for example, brachiation and bipedal ground walking. However, these monkeys have taken a very different path than the hominoids because, unlike us, they have an extraordinarily well-developed tail, which they use to hang from branches.

What if we were to be able to somehow see *Morotopithecus bishopi*, perhaps carrying an infant and walking bipedally along the ground? I suspect that any one of us would know instinctively and immediately that we were seeing a creature that was closely linked to who we are and how we became human.

Evidence From the Shoulder Joint and Femur of *Morotopithecus*

In 2004, MacLatchy reconfirmed that one of the fossils found in Moroto is indeed a portion of a scapula of *Morotopithecus*. The features on this bone demonstrate that the ape from which it came was capable of the types of overhead arm use that distinguish the living apes from the monkeys. Based on the vertebra and scapula, upright posture seems certain. MacLatchy accepts a quadrupedal posture for *Morotopithecus*, but there is no specific evidence for this. She does accept that the femur is modified in a variety of ways from those found in monkeys. Some of these modifications involve features she relates to vertical climbing. However, adaptation and optimizations for vertical climbing in the femur are essentially indistinguishable from optimizations for bipedal walking. Because all of the hominiform hominoids that are descendants of *Morotopithecus* demonstrate considerable skill and specialization for bipedal walking, it is entirely reasonable to expect that the skeletal modifications of *Morotopithecus* reflect bipedal walking in this common ancestor (or sister group) of the hominiforms.

After *Morotopithecus*: Size and Ground Living

Following the spinal innovation of *Morotopithecus bishopi*, we have three helpful touchstones in the range of 12 to 15 million years ago that help us understand the watershed events among the hominoids. One is the emergence of a lineage that leads to *Oreopithecus*, a European ape that had a long flexible spine and appears to have been bipedal

(Moya-Sola et al. 1999, Rook et al. 1999). The second event is the emergence of a lineage that leads to *Pierolapithecus catalaunicus* (Moya-Sola et al. 2004), an upright species with hominiform lumbar vertebrae. The third event is the divergence of the orangutan, a large-bodied ape now found in Asia that engages in bipedalism on the ground and uses all four limbs to clamber about the trees in a fashion often called *quadrumanual*. In a variety of models of human evolution, this is taken as the point at which body size increased very significantly, making full-time life in the trees difficult. It is therefore suggested that at this point, a major role for life on the ground commenced. This required anatomical innovations for successful locomotion on the ground as well as in the trees.

Here again, it is worthwhile to pause and reflect on the larger narrative. At each branch point after *Morotopithecus*, we have either an extinct lineage, or a living descendant that is an accomplished and habitual bipedal walker, but no quadrupeds. Why is it impossible that human bipedalism relates in a direct way to all of the bipedality apparent in every thread of its ancestral lineage?

Evidence From the Footfalls

One of the more interesting innovations seen in the bipedal locomotion of some hominoids has been termed "heel-strike" walking (technically *heel-strike plantigrady*). In many quadrupedal animals such as cats, dogs, and monkeys, each impact of the hand or foot upon the ground is started by contact of the fingers or toes, or by the middle of the hand or foot. In human, chimpanzee, and gorilla walking, the heel strikes first. Anatomically, this seems to be reflected in the fairly massive *calcaneus* (heel bone) seen in the African great apes and man. Gebo comments on this in the context of terrestrial quadrupedalism (1996); however, it seems that "heel-strike" walking makes the most sense as a component of bipedalism on the ground (Vereecke et al. 2005). The idea is that when walking bipedally on a branch, an ape will most likely want to use its opposable big toe to help grasp the branch as the footfall starts. Hence, a mid-foot impact or a toe-oriented (*digitigrade*) footfall seems safest. However, on the ground, there seems to be an advantage to impacting with the heel because it keeps the toes from catching on anything on the ground, and prevents the toes from becoming unduly stressed.

When gibbons walk on branches and vines, they take advantage of their opposable grasping big toes. Each step begins with the contact of the mid-foot and the grasping front of the foot. In terrestrial walking, however, the great apes and humans start with the heel and then roll the foot forward toward the ball of the foot for the propulsive push-off. Gibbons walking bipedally on the ground impact their midfoot along with the heel (Vereecke et al. 2005). Although this is certainly subject to various interpretations, it may be that heel-strike bipedalism developed early in the history of the hominiform hominoids, but is less frequently used and has fewer persisting foot adaptations in the primarily arboreal hylobatids and orangutans.

Going forward from this point in time, we have the divergence of three separate lineages: one for the gorillas (split off at 8–9 million years ago); one for the chimpanzees (split off at 5–6 million years ago); and one for the hominines (the dozens of species, including *Australopithecus* and *Homo*, that have humans, but no apes, as their descendants). The events in this tripartite story of gorilla, chimpanzee, and human are the subject of the following chapter.

Conclusion

It is a widely held position among anthropologists who study *Australopithecus* and *Homo* that upright bipedalism must have arisen from a quadrupedal common ancestor with the chimpanzee. However, there is no definitive fossil evidence at all for this position. All the evidence, which is reviewed in detail in the following chapter, actually appears to point the other way when it is fully and fairly considered. Certainly, some theoretical models and historical views exist that predate our fossil discoveries and thus support a quadrupedal ancestor for the so-called hominine lineage of human ancestors. Inferences have also been made based on aspects of anatomy that do not bear specifically on posture and locomotion. However, the evidence from the fossil record and the evidence from our understanding of lineage relationships has failed to support these positions in any definitive way. The time has come for the specialists involved to sagely put aside four decades of hypotheses based solely on the quadrupedal ancestor.

Granted, these views are held by a very large majority of trained paleoanthropologists. And who knows, they may be vindicated in the end. However, there is clearly absolutely no basis for insisting that that a primarily upright bipedal last common ancestor could not have been present at virtually every branch point that leads away from our own hominiform lineage. The fact is that a quadrupedal ancestor for the hominines remains to be proven. This is not a matter of lineages, divergence times, and relationships; rather, it is a matter of the functional anatomy of actual musculoskeletal systems and the fossilized skeletal residua of ancestral forms. All of these pieces of evidence must be carefully evaluated in the light of their various possible forms in either bipedal or quadrupedal ancestors.

There is just no way to deny the mounting evidence from anatomy, genetics, embryology, and paleontology. Almost certainly, the ancestor shared by the quadrupedal apes and the bipedal humans was itself an upright biped. In fact, two examples of this upright ancestor for chimpanzees and humans may have already been found. These are the fossil hominoids *Sahelanthropus tchadensis* and *Orrorin tugenensis*. The problem with these two is that they seem to have lived 6 to 7 million years ago, which is before the time predicted for the chimp-human split (6 million years ago). Yet, it is now accepted by a growing majority of paleoanthropologists that *Orrorin* was primarily bipedal, based on evidence from the femur (Wood et al. 2002, Ohman et al. 2005). The case for *Sahelanthropus* as an upright biped is less widely accepted, and is based on the orientation of the connection between the skull and the spine of the neck (Zollikofer et al 2005). Nonetheless, because these two species were bipeds, most scientists assume that the human-chimp split must have taken place before they lived. Analysis of the cranial shape of *Sahelanthropus* shows it as more similar to *Australopithecus* than to modern chimps and gorillas (Guy et al. 2005). However, if it is compared only to modern humans, chimps, and gorillas, it comes out much closer to the apes.

The problem here is the assumption that the common ancestor of humans and chimpanzees was not a biped. If that were the case, we could use bipedalism as the deciding feature for including a specimen in the hominine subfamily and excluding chimpanzee ancestry. However, I do not believe the evidence supports this. The common ancestor of chimpanzees and humans was most likely bipedal, and both *Sahelanthropus* and *Orrorin* fit the bill for pre-split common ancestors of chimps and humans.

Using a recent definition of "human" provided in a 2004 article in *Nature* by David Begun, a University of Toronto anthropologist, both *Sahelanthropus* and *Orrorin* fit the description of being human. However, both may be reasonable candidates for ancestry of the chimpanzee lineage as well, because they seem to have lived before the date of the human-chimpanzee split predicted by biochemistry. Therefore, if the evidence for an upright, bipedal, hominiform lineage is accepted, then either of these two species can be seen as a human ancestor for the chimpanzee. Our definition of what is "human" may need to be adjusted in some way, but this will not be the first time questions about humanity have been raised with regard to the chimpanzee (Tuttle 2006a).

The Four Great Hominiforms

A Secondary Return to All Fours—The History of Knuckle-Walking

On December 31, 1853, the famous vertebrate anatomist Sir Richard Owen (who was the first to describe the great apes as "knuckle-walkers") and the artist-sculptor Waterhouse Hawkins hosted a dinner *inside* the mold of a full-sized statue of an iguanadon. The dinner took place at the newly built Crystal Palace north of London, the first steel and glass building ever assembled. Its developer had commissioned Waterhouse Hawkins to build a series of 33 full-sized dinosaurs, the natural wonder of that age, and perhaps of this age as well. Hawkins had built his first full-sized dinosaur replica for the London World's Fair two years earlier in 1851.

For Owen, it was part of his relentless promotion of his newly coined word *dinosaur*. He was still agitating for government funding to build a Natural History Museum in London, and he saw the public spectacle of the dinosaurs as a critical component of his publicity campaign. He was popular with Queen Victoria and Prince Albert, both of whom often attended his lectures. However, the dinner for 20 scientists inside a dinosaur was his greatest public relations triumph. It was reported on with drawings in the major papers, and is still the most widely remembered moment of his long and many-storied life.

He first published his new word in 1842 (Owen 1842). He described the etymology as from the Greek *deino* for fearsome, and *saur* for lizard. Without his wonderful new word, he would have a collection of bones from an alphabet soup of scores of large, long-legged, extinct lizards; instead, he had the fearsome symbol of the dinosaur. After 11 years of relentless promotion, the dinner at the Crystal Palace was the cornerstone of his effort. After this, he could be quite certain that no one would ever forget his new word.

The Origin of the Knuckle-Walking Concept

Sir Richard Owen truly understood the latent intellectual power that could be commanded by the creation of a new word. Two of his most enduring achievements in this area are his invention of the word dinosaur, and his definition of the term *homology* (see Chapter 5). Beyond this, he is well-remembered as a most prominent early colleague and mentor of Darwin, who later became the evolutionist's chief antagonist. In 1859, the year of publication of Darwin's *The Origin of Species*, Sir Richard Owen published his own major work on the classification of mammals (Owen 1859). In this work, he characterized the apes as belonging to two groups: brachiators, including the arm-swinging gibbons; and knuckle-walkers, including the orangutan, the chimpanzee, and the newly discovered species called the gorilla (which Owen formally named and described as well).

All three of Sir Richard Owen's most famous words are now facing challenges. From a cladistic point of view, there is actually no such thing as a dinosaur. The study of morphogenetics has shown that homology is only an approximation of some results of the body-forming genes. And the term *knuckle-walkers* is now proving to be a very misleading term. Of the three original knuckle-walkers, we now understand that orangutans generally do not knuckle-walk, which leaves just the chimpanzee and gorilla. Despite intensive search for 150 years, there are still no completely convincing anatomical adaptations in the bones of the hand or wrist of chimps and gorillas that seem to indicate knuckle-walking. What is evident upon examination and reflection is that the three great apes have a similarity in posture that I call *diagonograde,* or diagonal posture, and that the pattern of hand use should not be the focus. Once this is understood, a complex suite of major adaptations of the bones is revealed that adapts each of these apes for its unique type of locomotion. Most importantly, although there are similarities, these three great apes represent three quite independent and parallel evolutionary developments.

Knuckle-walking in chimps and gorillas

Monkeys run on the ground using all four limbs, similar to dogs and cats. Typically, their hands meet the ground in the same way as a human's when assuming a quadrupedal posture (on hands and knees); the palm of the hand is flat on the ground (*palmigrade*). This position is also used by gibbons and orangutans. Baboons, on the other hand, stand on their fingers—a posture called *digitigrade* (see Figure 10-1), which is also used occasionally by humans. The hand can also be closed into a fist, so that the ground is impacted either with the first knuckle or with the curled small finger. Fist-walking on the first knuckle is the most common type of hand-to-ground impact used by orangutans, and is quite common for chimpanzees.

Gorillas and chimpanzees are arm-swingers in the trees, although not nearly as agile as the orangutans, gibbons, or siamangs. On the ground, all of the apes walk bipedally. However, for most of their movements on the ground, the gorilla and chimp use an unusual walking position in which the elbow is kept straight and the second knuckle meets the ground (see Figure 10-1). Modern humans cannot do this because it requires the first knuckle to be bent backward beyond what the hand is capable of when the second knuckle is flexed. It is this uniqueness—this complex hand use by our close relatives that we cannot quite match—that has drawn all the attention.

In gorillas, this usually means contacting all four non-thumb fingers (digits two, three, four, and five) with the ground, and keeping the hand perpendicular to the direction of travel. In chimpanzees, this is typically done by impacting the third and fourth digits, most commonly parallel or oblique to the direction of travel.

Errors in the Wrist

Much of what is now widely understood about knuckle-walking comes from the work of Russell Tuttle, the University of Chicago physical anthropologist who was the advisor for my master's thesis (Tuttle 1967, 1969, 1974, 1975, 1983), and the quite separate work of Farish A. Jenkins, Jr. (the Harvard University anatomist and paleontologist who was also one of my Ph.D. thesis advisors) and John Fleagle, the Stony Brook primatologist (Jenkins and Fleagle 1975).

Sadly, a relentless series of brilliantly detailed reports by Owen J. Lewis, a University of London anatomist based at the ancient and fabled St. Bartholomew's Hospital Medical College, have virtually demolished the idea of knuckle-walking as a distinct, detectable anatomical capability of the hands of the great apes (Lewis 1971, 1977, 1985). Culminating in a magisterial, 368-page, Oxford University Press book on the subject, Lewis has not minced words:

> Jenkins and Fleagle...had no appreciation of the complex nature of the motions...because of an apparent misunderstanding of the true mechanics of the movements.... Their chain of reasoning, resting upon surprising misinterpretations of published data...[and] other misinterpretations compounded the confusion. The studies cited above which played down the likely role of suspensory posturing in hominoid evolution, although based on inadequate data and flawed interpretations, have had the unfortunate consequence that they forcibly imply that the ancestral apes must have been essentially "dental apes"—ape-like in dentition but quadrupedal and monkey-like postcranially (Lewis 1989, 72, 87–88).

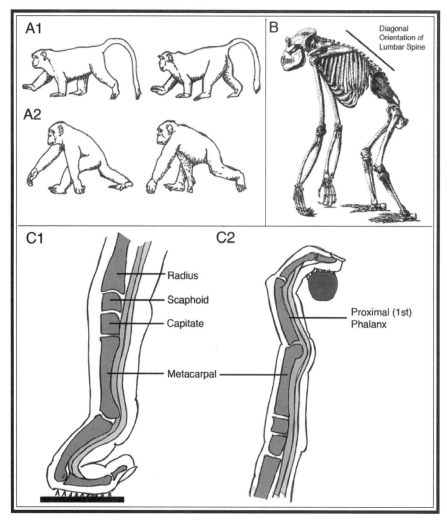

Figure 10-1: Hands and spine in primate walking.
A1) A monkey walking with palmigrade/plantigrade posture, with the spine parallel to the ground (pronograde).
A2) A chimpanzee knuckle-walks using the second knuckle (phalanx) for ground contact. The spine is diagonal to the ground (diagonograde).
B) A gorilla skeleton demonstrating knuckle-walking posture with diagonograde orientation of the spine.
C1) Biomechanics and anatomy of knuckle-walking hand posture. Note the joint between the first knuckle and the long bones of the hand (metacarpo-phalangeal joint).
C2) The hand in suspension. Bony ridges limit extension of the joint, showing that similar mechanical demands arise in suspension and in knuckle-walking.

Tuttle has pointed out that Sir Richard Owen got it wrong in that orangutans typically do not knuckle-walk (1975). Rather, they prefer either bipedal walking or quadrupedal use of their hands in a fist formation. Tuttle also accepted that Lewis (1972, 1973) disproved earlier ideas on "adaptations" of the wrists of gorillas and chimps for knuckle-walking. This is actually a huge muddle. Beginning with Owen, heavily promoted more recently by Washburn, and then proceeding on through Jenkins, Fleagle, and the Stony Brook group and their students today, the role of knuckle-walking as a predecessor stage in human evolution has made a totally indelible impact. This is not only academic but cultural; everyone seems to accept this. But the experts here are Jenkins, Tuttle, and Lewis. Unquestionably, Lewis wins and settles—there do not seem to be any adaptations in the hand or wrist for knuckle-walking. But this is a bit of a problem: The key linchpin of nearly all of the modern theories of human evolution is effectively a no-show. What is going on here?

In fact, there are several locomotor specializations in the hominoid great apes. It is Sir Richard Owen's term *knuckle-walking* that is the real problem. Yes, chimps and gorillas do place their weight on the middle phalanx (the middle of the three parts of the finger) when they walk quadrupedally. However, this draws attention to a minor or trivial aspect of what they are doing. As Tuttle has pointed out in accepting Lewis's criticisms (1975), both orangutans and humans can perform a version of knuckle-walking reasonably well when they want to, even though they cannot be described as primary knuckle-walkers.

Lewis (1989) draws attention to the fact that Jenkins and Fleagle, in their very widely read and referenced 1975 article on knuckle-walking, experienced a frustrating failure of their scientific method. In a variety of studies done at the laboratory at the Museum of Comparative Zoology at Harvard we used high-speed motion-picture X-rays (cinefluorography) to analyze the movements of bones by observing animals walking on treadmills, swinging on ropemills, and so on. This is intended to show how an animal actually moves, and provides a fantastic window into functional anatomy in some settings.

Although some joints are easy to track in cine-X-ray, the wrist proved to be an undecipherable mess in the knuckle-walking chimp. This is because there are many bones involved whose images overlap in a lateral, two-dimensional image, as well as side-to-side movements that affect the apparent size of each bone and joint as they move closer to or away from the X-ray camera. Moreover, some of the critical movements are small. Because of this, Jenkins and Fleagle had to supplement their results by taking ordinary still X-rays of ape hands (the scientists moved the hands into various positions while the apes were under anesthesia). This is a method that has been used to study primate

hands for more than 100 years (Bryce 1897), but has none of the special in-
formational quality provided by the cine-fluoroscopy of a conscious, naturally
behaving animal.

Not only are the movements described in the Jenkins and Fleagle 1975
paper a problem, but the key piece of anatomy is apparently wrong. Lewis
points out that another 1975 major review of primate locomotion and
knuckle-walking by another Stony Brook scientist, Jack Stern, bases a major
theory of hominoid evolution upon the Jenkins and Fleagle study. Lewis
writes:

> In this survey he (Stern) is influenced by the findings of Jenkins and
> Fleagle (1975) and repeats the fallacy introduced by these authors
> that the African apes have a "centrale" facet on the dorsum of the
> capitate and that this might be considered an adaptation to
> pronograde quadrupedalism…. Much of the basis for the…hypothesis
> is flawed (Lewis 1989).

Theoretical Distortions From the Wrist Muddle

When three major figures in an academic field report and cross-reference each other on incorrect information that is at
the heart of a major and widely accepted theory, and then produce textbooks,
give keynote addresses at major meetings, review grants, and so on, the consequences are painful to contemplate. Lewis has tracked this and finds no public
withdrawal of the error (1989). Even today, 30 years later, most specialist
anthropologists take it as fact that there is a major shared adaptation of the
hominoid wrist in gorillas, chimps, and humans that proves these three groups
share a common, quadrupedal, knuckle-walking phase.

Brian Richmond, a University of Illinois–Urbana anthropologist, and
David Strait of the New York College of Osteopathic Medicine published a
very prominent report in *Nature* in 2000 that received enormous attention
(Richmond and Strait 2000). However, this proved to be little more than a re-
reporting of a feature originally described by Tuttle in 1969, and dismissed as a
knuckle-walking adaptation by Lewis in 1989. This feature is a ridge and groove
at the joint between the radius and the scaphoid bone of the wrist (see Figure
10-1). Richmond and Strait pointed out that this feature occurred in the
wrists of two of the earliest species of *Australopithecines*. In their widely quoted
2000 *Nature* paper, they showed images they described as "video," which demonstrated how these features engage during weight-bearing. They start back-
ing away from the claims immediately in their 2001 article (Richmond et al.
2001), in which they cite the 1975 Jenkins and Fleagle paper 16 times over 30
pages without detailing Lewis's objections. They show the same wrist X-rays,

but add the detail that these are from deceased apes manipulated for the X-ray, thus avoiding any confusion that their data in the *Nature* paper might have been gleaned from live animals. Other specialists (Dainton 2001, Lovejoy et al. 2001, Preuschoft 2004) point out that the evidence provided by Richmond and Strait is unconvincing because the limitation of wrist extension they describe may be related to climbing and suspension activities rather than to knuckle-walking, and because their analysis of wrist function is significantly flawed (see Figure 10-1).

In the next paper by Richmond on this subject, he backs away further from his own claims. Regarding adaptations in the middle of the wrist, he states that the data provide "weak" and "mixed" support for the hypotheses (Richmond 2006, 112, 115). Essentially, efforts to confirm limited wrist extension in living, anesthetized primates proved to be incomplete and contradictory. Richmond tested two chimpanzees and found limited extension (30 degrees), but Tuttle had tested 17 chimpanzees under similar conditions and found the mean to be 42 degrees, with considerable variation (1969). Tuttle's 10 gorillas averaged 58 degrees, which is comparable to the wrists of monkeys that do not have the features of wrist anatomy discussed in the Richmond and Strait 2000 article. Again, the 1975 Jenkins and Fleagle article is cited as the source of information 10 times, with no comment on the objections from Lewis in 1989 that dismiss the results of that Jenkins and Fleagle article.

Time and again, the anatomical evidence proposed to support the hypothesis that human ancestors walked on their knuckles has proven to be based on error. No such anatomical characters appear to exist. As Lewis has repeatedly stated, and as Tuttle has accepted, there are no shared anatomical features of knuckle-walking to be discovered in the wrist of humans, chimps, or gorillas. Tuttle, despite his extraordinary level of expertise regarding knuckle-walking, and a 37-year record of publications on the subject, has never reversed his long-standing position that there probably was not a knuckle-walking phase in human evolution. Recent revisits to well-known traits have fit in with ideas that much of the anthropological community desperately wants to find validation for. As much as these papers have stirred debate, they have added no solid evidence.

Results reported in a 1999 paper by Mike Dainton and Gabrielle Macho, both anatomists from the University of Liverpool, reveal another major issue. In the work by Richmond and Strait, a small number of traits in the wrist are analyzed to show the special affinity of the bones of the early *Australopithecines* with those of chimps and gorillas. Dainton and Macho analyzed a much larger number of features and produced a very different result. The wrists of chimpanzee and gorilla were quite different from each other in a way that strongly supports the independent, parallel evolution of two different forms of knuckle-walking, and does not support the association with the wrist in *Australopithecus* (Dainton 2001).

For many years, our models of ape and human lineages suggested that chimpanzees and gorillas shared a common ancestor after separating from the human lineage. However, Maryellen Ruvolo, the Harvard University anthropologist and molecular biologist, has provided a series of proofs (Ruvolo et al. 1991, Ruvolo 1994, 1997, 2004)—now widely supported by others—that it is human and chimpanzees that shared a common ancestor after the divergence of gorillas. This structure of the family tree impacts current discussions of knuckle-walking to a considerable degree. When Richmond and Strait produced their 2000 and 2001 reports, they provided lengthy and extensive reviews of models of the origins of human bipedalism, and conclude that there must have been a knuckle-walking phase. However, their argument is not really anatomical at heart. They assert that both chimps and gorillas knuckle-walk, and so, given the Ruvolo pattern of the family tree, it makes sense that the human ancestor must have knuckle-walked as well. This is not viable reasoning if knuckle-walking draws attention to a marginal aspect of how these apes locomote, particularly if what they are doing is merely similar, but independently evolved. This is a critical point because it relieves anthropologists from a perceived obligation to show that human ancestors passed through a knuckle-walking stage. If there is any confusion on this point, it should be clear that there is little or no actual convincing evidence here to support the frequently proposed role for knuckle-walking in human evolution.

Evidence From the Wrist

In his 1996 overview and in subsequent works, Gebo has stuck with the troglodytian model championed by Sherwood Washburn. His main supporting point for this is the anatomy of the wrist. He points out that a number of anatomists have drawn attention to the modifications of the wrist in the African Apes and man that, when compared to the wrist of other hominoids, appear to be designed for weight-bearing across the wrist. He joins others in arguing that this is the proof that the common ancestor of humans and the African apes was a quadrupedal knuckle-walker.

The influence of brachiation on human anatomy is easy to accept. There is a good argument that arm-swinging or at least overhead suspension when in trees has played an important role for humans throughout our evolution from the time of *Morotopithecus*. Humans are perfectly capable of arm-swinging, even as adults. The fact is that the knuckle-walking argument for the basis of increased surface area and slightly decreased mobility in the wrist cannot serve as the only reliable basis for proving that a knuckle-walking phase existed somewhere in the course of human evolution. This is particularly true because the anatomical case for this is not accepted by leading experts in the area of hominoid wrist anatomy. When we look at the skull, neck, shoulders, arms, internal organs, chest, lumbar region, pelvis, legs, and feet, all we see is evidence compatible with upright posture and bipedalism, more or less. Only the wrist has seemed to prove the existence of a troglodytian quadrupedal phase, but this is in no way a compelling bit of evidence. If the strong wrist is the only support for this theoretical edifice of ancestral human knuckle-walking, then the collapse of that theory is likely.

Russell Tuttle argued that knuckle-walking is a very specialized type of locomotion, requiring an unusual set of anatomical features that are unlikely to evolve multiple times (Tuttle 1969). However, Tuttle later acknowledged that orangutans, which lack these modifications, are able to knuckle-walk perfectly well under various circumstances (Tuttle 1975). His most impressive explanation of the fundamental locomotor biology of knuckle-walking (Tuttle 1969) seems to be missed by nearly every scientist who discusses it. Chimps and gorillas both do something extremely unusual in their quadrupedal gait: They keep their elbows fully extended through the gait. In most mammals, the elbow is flexed when the fore limb hits the ground. This allows the animal to use a powerful contraction of the triceps to generate locomotor drive. In addition, the wrist is extended so that the hand flexors can add a powerful propulsive stroke as well. Neither of these motions are possible when a straight elbow and straight wrist posture is used. Indeed, when Tuttle used electronic techniques to monitor the triceps of chimpanzees during fist walking and knuckle-walking, they made little or no use of the triceps (Tuttle 1983). Tuttle argues that in knuckle-walking, the hyperextension of the first knuckle allows the ape to powerfully flex the fingers to add force to the stride (Tuttle 1969). This is similar to the use of the arms in brachiation—a straight elbow and wrist, with active flexion of the fingers.

The straight elbow/straight wrist posture is also used by great apes when fist-walking. A key feature of this posture that has not previously received significant attention is that it keeps the shoulder high relative to the hips. In fact, because the knee is usually flexed throughout the entire stride, the result is a diagonal posture of the torso (see Figure 10-1). Rather than progressing with the body parallel to the ground, as in nearly all other mammalian quadrupeds, these apes—the orangutans, chimps, and gorillas—either walk fully upright bipedally, or walk on all fours with their body at a diagonal angle of about 45 degrees from the ground. There is powerful evidence that this diagonal posture of the spine is the actual substantive component that makes the locomotor behavior of these three separate ape lineages appear to be so similar, and that the hand postures are really secondary and variable.

Is "Knuckle-Walking" Really a Spinal Adaptation?

Posture at an Angle

There is a critical mechanical reason why chimps and gorillas position the spine at an angle that is typically greater than 45 degrees, rather than horizontally, when they progress on four limbs (see Figure 10-1). This mechanical requirement relates to the unique anatomy of the hominiform lumbar spine. By effectively lengthening the fore limb relative to the leg, knuckle-walking produces a more vertical spine orientation and a steeper diagonal. Because of the length of the hand in a gorilla, the effect of the straight wrist is considerable.

The *Morotopithecus*/human modification of the lumbar spine is truly effective only in creatures that are fully upright. The whole system works very badly for anything resembling *pronograde* (horizontal body posture) locomotion. Recall that the fundamental mammalian support mechanisms for the lumbar spine are missing in the basic hominiform lumbar spine that first appeared in *Morotopithecus* (see Figure 9-6). The knuckle-walking posture is more *diagonograde* (held somewhere between upright orthograde and horizontal pronograde). Indeed, in order to accommodate the locomotor style of chimpanzees and gorillas, despite the diagonal body posture, the hominoid lumbar spine undergoes some further dramatic modifications added on to the hominiform pattern of *Morotopithecus* and *Homo*.

Orthograde, Pronograde, and Diagonograde

What is really unique about the method of locomotion employed by orangutans, chimpanzees, and gorillas is the spinal adaptation. With very few exceptions, mammals carry their spines either parallel to the ground (pronograde) or perpendicular to the ground (orthograde). Jack Stern of the Stony Brook group has proposed the term *anti-pronograde* to include primate postures in which the body is held at an angle greater than 45 degrees. This would include vertically oriented slow climbers such as the prosimian loris, and various vertical clingers and leapers among the prosimians, brachiating gibbons, and bipedal humans. A more specialized term is needed to describe the unique pattern of locomotion employed by chimps and gorillas, and diagonograde is a good descriptor.

A striking aspect of quadrupedal locomotion of the great apes is the diagonal angle of the spine. The Moroto modification that the great apes share with humans is a horribly ineffective structure for horizontal posture. There is very little mechanical resistance to the collapsing force of gravity other than what is provided by the vigorous continuous contraction of the back muscles, and nature seems to abhor the use of continuous high levels of energy just to stand still.

Design for Weight-Bearing in the Lumbar Vertebral Bodies

One feature of the basic hominiform lumbar vertebral architecture is that weight-bearing must pass through the lumbar vertebral bodies and disks during any type of movement or posture in which the hind limbs are used, except when brachiating or hanging by the arms. When weight is applied only to the posterior elements (the lamina and facets), then the whole weight of the creature is borne by the stretched ligaments of the small facet joints—a condition that is not only painful and destructive, but also goes beyond the possible mechanical design constraints of the small, flexible ligaments. Because these facet ligaments at the various level of the spine are "in series" with each other (stacked one upon another), they may fail at multiple spinal levels if each one is forced to carry much of the entire weight of the body.

Some mammals, such as the ungulates, do rely on the posterior elements for weight-bearing when the spine is extended (Filler 1986, 2007a). However, the *Morotopithecus* and the *Homo* lumbar vertebra cannot be employed in this fashion. Mammals that use their lumbar posterior elements to bear compressive weight typically employ bone-to-bone contact to carry the compressive forces that resist spinal hyperextension in horizontal standing and locomotion. In other creatures as varied as the opossum, the elephant, and the rhinoceros, the entire weight can be borne on the posterior elements because of a massive series of bone-to-bone, weight-bearing surfaces in or near the facet joints of the posterior part of the vertebra (Filler 1986, 2007a).

Figure 9-6 shows how many other mammals, such as the non-hominoid primates, the carnivores, and the rodents, handle their lumbar spinal weight transmission when in a pronograde posture that causes spinal extension. The lumbar transverse processes become loaded in extension, and drive the vertebral bodies together at the level of connection of the LTP to the vertebral body (see also Figure 9-2, and Filler 1986, 2007a). If any weight-bearing needs to be passed to the posterior elements, this can be handled by bone contact involving the facet and the bone at the base of the styloid. However, both of these extension-resisting capabilities available to most primates are absent in *Morotopithecus* and *Homo*, and they have no mechanism for applying posterior element, bone-to-bone contact as in the ungulate spine. With the basic hominiform lumbar architecture, the weight must be carried by the anterior elements (the disks and vertebral bodies). This occurs automatically when a hominiform stands upright. The weight remains on the vertebral body and disk as the hominiform bends forward to about 45 degrees; past that point, however, in a chimpanzee or gorilla, the hands typically reach the ground, and a quadrupedal stance commences. Gravity now produces extension of the spine. The forces are similar to what is shown in Figure 9-6, but the basic hominiform lumbar spine lacks the ligamentous and bony structures to handle the resulting forces comfortably.

In humans, and in *Morotopithecus*, there is no means of keeping the force of weight-bearing out of the facet ligaments when the spine passes into extension in a pronograde position. This is particularly true because hominiforms do not use an arched or flexed lumbar spine when walking or standing, as in the type of mammalian flexion shown in Figure 9-6 in the previous chapter. Humans try to avoid these postures, and use them only very rarely. When gorillas and chimpanzees are observed in quadrupedal stance, they also tend to avoid strictly pronograde posture. By keeping their elbows straight, wrists extended, and knees flexed, gorillas and chimps tend to keep their torso at or above 45 degrees (see Figure 10-1).

Locking and Blocking of Facets

In orangutans and gorillas, there are independent but similar modifications that allow the posterior elements to bear weight when the body axis passes below 45 degrees. These are bone-to-bone, facet contacts that mimic what is seen in ungulates (Filler 1981, 1986, 2007a), but that are never seen in humans (see Figure 10-2). In fact, locking facets do not occur in any other primate, rodent, or other members of the ancient euarchontogliran superorder to which these types of animals belong. As the weight-bearing passes to the posterior elements in these apes, bone-to-bone engagement occurs at the facet locks. Therefore, the facet ligaments never bear the full weight of the body. Interestingly, in orangutans, the facet modification occurs in a plane parallel to the skin of the back and involves a structure on the pedicle (pedicle facet lock); in gorillas it occurs in a plane perpendicular to the skin of the back, and involves a structure on the lamina (laminar facet block). This is fairly clear evidence that these two modifications developed independently in response to similar adaptive pressures in the two groups. This is not an adult degenerative change, as they are readily seen in juvenile individuals in orangutans and African apes (Filler 1986, 2007a).

Iliac Suspension of the Chimpanzee Lumbar Vertebrae

In chimpanzees, there are no locking facets. In this species, the origin of the longissimus back muscles is repositioned along with the iliac crest to be in the same plane as the lumbar transverse processes (LTPs). Unlike gorillas, humans, and other Moroto-type apes (hominiforms), chimpanzees have a flat, bladelike LTP that reflects the specialized ilio-lumbar ligament complex that helps them rigidify the lumbar spine. In addition, the longissimus muscle attachment undergoes further modification, and is shifted to the base of the LTP and partially onto the mamillary process on the side of the facet joint itself (Filler 1979).

Figure 10-2: Additional modification of great ape lumbar vertebrae.

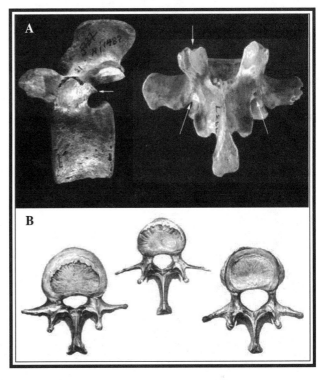

A) Orangutan facet locks and gorilla laminar blocks. The orangutan and the gorilla have independently evolved different mechanical blocks that limit extension of the lumbar spine and prevent hyperextension in quadrupedal, diagonograde posture. A (*left*) Lateral view of orangutan lumbar vertebra. The inferior facet is close to the pedicle (compare with human configuration in Fig. 9-2) and a locking extension assures hard bone to bone contact with the superior facet of the next lower vertebra.

A (*right*) Dorsal view of gorilla vertebra showing the groove on the superior facet and notch in the lamina that that limit extension.

B) Comparative superior view of gorilla, chimp, and human vertebrae. Notice the thin, flat lumbar transverse process in the chimpanzee.

Part of this specialization in the chimpanzee involves the extreme upward elongation of the iliac crests relative to the lumbo-sacral transition. In a human with four lumbar vertebrae, all four vertebrae are typically above the iliac crest. In a chimpanzee, there is usually just one lumbar vertebra above the iliac crest (see Figures 10-3 and 10-4). This is why a human with four lumbar vertebrae still has a relatively long flexible spine while a chimpanzee with four lumbar vertebrae has virtually no flexible lumbar region at all (see Figure 10-3).

Additive Modifications of the Great Ape Lumbar Spine

The lumbar vertebral situation in the great apes represents another characteristic of evolutionary change. These animals achieve a useful anatomic arrangement by adding onto or modifying a feature that evolved for the benefit of an ancestor. In this case, the orangutans, chimps, and gorillas move beyond the human lumbar form so that they can be better adapted for life in the trees. There is no need for locking facets if the older primate lumbar configuration of ventrally projecting spokes in an elastic sheet is maintained. However, in the great apes that carry the Moroto modification in their lumbar vertebrae, a new type of mechanical support is needed if they will not be full-time upright bipeds. The locking facets of orangutans, the facet blocks of gorillas, and the lumbo-iliac suspension of chimpanzees each play that role of a parallel, independently evolved additive modification to the hominiform lumbar spine. Humans retain upright bipedalism and keep the original unmodified hominiform lumbar vertebral configuration first seen in *Morotopithecus*.

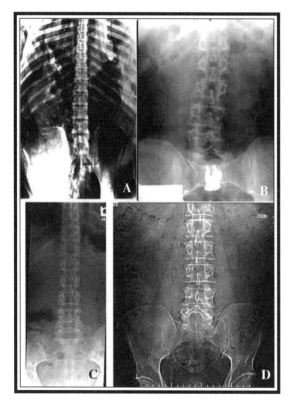

Figure 10-3: Lumbar X-rays in chimpanzee and humans. A) Adult chimpanzee spine demonstrating ultrashort, free lumbar region with four lumbar vertebrae. The ribs approach the iliac crests. L2, L3, and L4 are between the elongated iliac crests. B) Human with four lumbar vertebrae showing an effective, long, flexible lumbar spine despite decreased number of lumbar vertebrae. This individual has undergone an unintended and unproductive placement of an anterior surgical fusion plate between S1 and S2. C) D) Humans with six lumbar vertebrae including partial or "transitional" L1 bearing unilateral small rib on the right in (C).

Changes in Vertebral Numbers: Homeotics and the Hominoids

In addition to the mechanical transformation of the shape of the hominiform lumbar vertebra, there has been a change in the number of lumbar vertebrae in the lumbar spine. Most primates, including the monkeys, have seven lumbars. Carol Ward and colleagues (1993) were able to show that *Proconsul nyanzae* had six or seven lumbars, so it seems that the changes of lumbar spinal numbers may be restricted to the hominiform descendants of *Morotopithecus*. Overall, the upright hominiforms that are bipedal on the ground—the hylobatids, *Oreopithecus, Australopithecus*, and *Homo*—have five or six lumbar vertebrae.

Ultrashort spines have emerged in those hominoid descendants of the *Morotopithecus* clade—orangutans, gorillas, and chimpanzees—that engage in more horizontal types of locomotion. Across thousands of mammalian species, only gorillas, orangutans, chimpanzees, and two species of armadillos are known to have just 16 presacral vertebrae, thoracic plus lumbar (Filler 1986, 2007a). The standard number in mammals is 19, although some have more than 24 thoracic and lumbar vertebrae. The result in the great apes is common lumbar spine lengths of just three or four lumbars (see figures 10-3 and 10-4). The wings of the iliac crest of the pelvis meet the downward-angled ribs. The overall effect is that the flexible lumbar region is essentially eliminated entirely.

The Ultrashort Lumbar Configurations of the Great Apes

In all existing hominoids the tail has disappeared, and is replaced instead by a small remnant called the coccyx. The sacrum is always lengthened from the ancestral three vertebrae to include four or five vertebrae. The lumbar region in contrast is always shortened from the ancestral seven vertebrae, going down to between three and six vertebrae. In addition, the pelvis is mounted higher up the lumbar spine than in other mammals, and the sacrum is lengthened. A bigger sacrum is more suited to carry the full weight of a bipedal creature that does not use its arms for support when standing. The shortening of the lumbar region in species such as *Homo sapiens* may simply reflect the incorporation of the last lumbar vertebra into the sacrum to help this structure carry a larger share of the total body weight. However, as the shortening modification has progressed further in the great apes, it clearly brings the rigid pelvis very close to the rigid rib cage and greatly reduces the amount of flexibility in the lumbar spine (see figures 10-3 and 10-4). The brachiating New World monkey, the spider monkey *Ateles*, also has evolved a shortened lumbar region of just four lumbar vertebrae—another fascinating example of their parallel evolution with the hominoids.

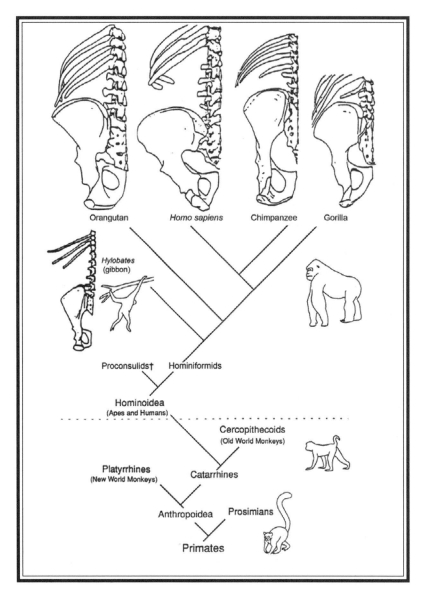

Figure 10-4: Hominoid spine phylogeny. Relationships of primate groups and hominoid species showing some aspects of difference in spinal anatomy. Notice the longer flexible lumbar spines in humans and hylobatids, contrasted with the ultra-short lumbar region in the orangutan, gorilla, and chimpanzee. The family name "hominiformid" is suggested here to describe a clade (natural group) that includes *Morotopithecus* and all existing apes and humans, as well as other extinct species whose ancestor had the human-style modification of the lumbar vertebra.

Basically, each of the hominoid species has its own unique lumbosacral configuration (see Figure 10-4). Clearly the genes that control development in this region are "in play," in evolutionary terms. In repetitive unchanging fashion, thousands of mammalian species are essentially identical in basic lumbosacral vertebral numbers. Not only are hominoids different from other mammals, but each hominoid species differs greatly from the other hominoid species in this region of the body.

The Nature of Spinal Number Changes: Homeotic and Meristic

The question of how vertebral numbers change has been a central question in genetics for more than 100 years. Fresh upon his participation in the rediscovery of Gregor Mendel's foundational work in the nature of heredity, William Bateson founded the modern field of genetics. Among his first great projects in the field was to seek to understand the way in which mutation affected heredity and evolution. Not only did his early work provide a foundation for genetics in the 20th century, but it also presaged the growing focal point of 21st-century genetics—namely, the nature of homeotic change and the genetics of the assembly of organisms.

In 1894, as part of this work, William Bateson provided a detailed account of the issue of variation in vertebral numbers as part of his classic foundational work, *Materials for the Study of Variation*. He provided the original definitions in that work for *meristic* change, in which the total number of vertebral segments might be changed, and for *homeotic* change, in which characteristics of a given vertebra might be changed (for instance, the 20th vertebra changing from a thoracic to a lumbar). However, he also provides a detailed account of the issue of lumbosacral anatomy in the apes and humans, and concludes that you cannot reliably state whether variation in thoracic, lumbar, and sacral numbers is accomplished by either of these two genetic methods. In other words, there's no way to know if the change is meristic (adding or removing a segment) or homeotic (changing the anatomical development or characteristic of a particular segment). The problem is that each vertebral segment may be defined in various ways, based on the position of nerve roots, the type of lateral process or rib, and/or pattern of joint position. Bateson even considers a most unusual gorilla individual that had eight cervicals, and shows that the problem still can not be resolved. He points out that all known variation of this type is at adjacent regional borders. If a rib appeared in isolation (non-adjacent) on the fourth lumbar with three normal lumbars above it, then that would clearly be a homeotic mutation. He also points out that comparison between a snake with 175 segments and a lizard with 75 segments can reveal a true meristic change, as numerous additional segments have been added.

Shifting Features Along the Spinal Segments

In a recent extensive review, Pilbeam (2004) assembled data derived from a wide variety of sources in order to assess the number counts of vertebrae in primates. Most of the data relies on the presence or absence of ribs, and participation in the fused sacrum to define cervical, thoracic, lumbar, sacral, and caudal. (For a discussion of these methods, see Haeusler et al. 2002). In my own primary data reporting (Filler 1986), and in my review of primate data (Filler 1993), I followed Sir Arthur Keith (1902) and Adolph Schultz (1969) by presenting an array of varying segmental phenomena, including facet orientation and a variety of other osteological features for the mammalian overview, as well as nerve plexus position for primates. Due to the advent of MR neurography imaging (Filler et al. 1993, Filler et al. 2004), it is now also possible to document nerve plexus variation in living humans (see Figure 10-5).

Because the mix of serially distributed characters that appear on any vertebrae may shift and change—often becoming separate from each other by more than two segments—I lean toward a homeotic explanation for the shifts. This means that genes that assign characteristics to a given vertebra in the spinal series may change their target. Meristic changes do take place in hominoid evolution, but this has only to do with tail loss. Some authors have suggested that vertebral counts in hominoids provide precise numbers of somites; count the cervical, thoracic, lumbar, sacral, and the coccygeal vertebrae, and you will have an exact number. In fact, a few additional somites form during human embryogenesis, but these are later subject to reabsorption by the developing body. Most interestingly, this meristic process seems to be more advanced in the African apes than in humans. The human embryo forms eight or nine caudal somites, although only four or five typically survive as coccygeal vertebrae. The numbers are fewer in the apes (Schultz 1969).

The Sequence of Events in Hominoid Lumbar Numbers

The fact that about 90 percent of humans have five or six lumbar vertebrae could have arisen in two ways. This may represent a relatively primitive condition with little change. Alternately, as Pilbeam has suggested, the number could have dropped from seven down to three or four (for a knuckle-walking phase), and then gone back up to five or six again for hominine bipedalism (Pilbeam 2004). He points out that it is not convincing to suggest that a segment was lost in the great apes and cannot be recovered, and I certainly agree with that position. What is at issue is the possibility that the shortening altered the overall function of the spine in such a way that it strongly engaged adaptation/natural selection. I believe that it is the diagonal posture of the great apes that has been served by ultrashortened spines.

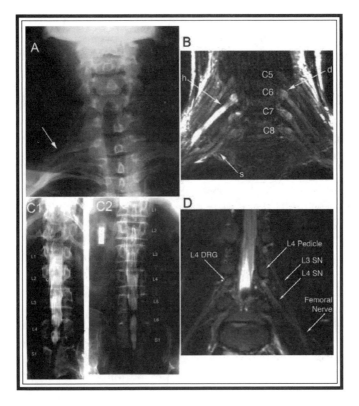

Figure 10-5: Homeotic variants in humans.
A) X-ray showing cervical rib at seventh cervical vertebra in human with nerve compression symptoms.
B) Magnetic resonance neurogram (MRN) showing distorted course of C8 and T1 spinal nerves.
C) Comparison of human X-rays in individuals with four and six lumbar vertebrae.
D) Use of MRN to identify and count spinal nerves as they combine to form plexus and distal nerve elements.

If the human lineage passed through a knuckle-walking phase as Pilbeam has considered, then the lineage may have undergone the shortening and subsequent lengthening he suggests. However, if there was always an upright ancestor reliant on bipedalism, then such significant shortening with subsequent reversal is unlikely or unimportant. As demonstrated in Figure 10-3, a human with four lumbars still has a long, flexible lumbar spine, whereas a chimpanzee with four lumbars has an extremely short lumbar spine, with essentially no flexibility at all. Hylobatids obtain a shortened lumbar spine primarily by the transformation of the first lumbar into a ribbed thoracic vertebra;

in humans, the sacrum is extended upward to sacralize the last lumbar. In orangutans, chimps, and gorillas, both of these processes take place. The lumbar region loses one vertebra to the thoracic region and one to the sacral region. In some individuals, the thoracic region takes over two lumbars, and in some the sacral region takes over two lumbars. High level of variability for these traits are displayed in all hominoid species, so presumably they are quite amenable to evolutionary change if natural selection comes into play. In hominoids, the reduction in the flexible lumbar region by transformation of segments into the more rigid thoracic and sacral configuration seems to change the lumbar spine to the point that it alters function. By way of contrast, consider the case of the two-toed sloth (*Bradypus*) and the three-toed sloth (*Choloepus*). One has 15 thoracics and the other has 23 thoracics, yet there are few major differences in locomotion or lifestyle. For these two closely related species, the variation in vertebral number appears to be purely driven by mutation, and seems unrelated to any obvious effects of natural selection.

The situation for hominoid lumbar regions is different. Because of the Moroto transformation in the LTP, styloids, muscle attachments, and ligament configuration, a long, flexible lumbar spine is not suitable for horizontal postures. Here we have a feature with high levels of variation among individuals in a species, for a character that is under heavy selection for fitness for locomotion, transporting infants, and running at speed. This is the classic "playground" for parallel or convergent evolution, and this is what appears to have happened here in the various great apes.

Encountering similar pressures to optimize a return to quadrupedal locomotion, each of the lineages of orangutan, gorilla, and chimpanzee individually made the minor shift of losing an additional one or two lumbar vertebrae. In addition to the number change, however, a variety of related changes occurred in these lineages that further stiffened and shortened the lumbar region, and those additional changes (vertical facet locks on the pedicle in orangutans, horizontal facet blocking on the lamina in gorillas, and iliolumbar suspension in chimpanzees) have little or no sign of homology among the three groups.

Evidence on Numbers From the Hominoid Lineages

The last common ancestor of monkeys, apes, and humans—existing about 24 million years ago—almost certainly had the standard primate condition of seven lumbars. Other hominoids outside the hominiform clade (*Proconsul*) continued to have six or seven lumbars 18 million years ago. *Nacholapithecus*, a later proconsulid species, apparently had six lumbar vertebrae (Ishida et al. 2004). When the lineage of the hylobatids split off at 17 million years ago, the typical number of five or six lumbars may be inferred by the status of the living descendants. The lineage of *Oreopithecus* split off around 15 million years ago, and these apes were upright bipeds with five lumbar vertebrae. The *Oreopithecus* lumbar vertebrae demonstrate the same kind of "wedging" that is the basis of our human lumbar curvature or *lordosis*. They also show gradual widening of the vertebrae as they progress toward the lowest lumbar, which suggests upright weight-bearing, and hence bipedalism (Rook et al. 1999).

Subsequently, each of the ape lineages demonstrates spinal reduction— either because there was reduction of lumbar numbers in a common ancestor, or because this reduction was accomplished independently, in parallel. This would have had to occur separately three times: in the orangutan lineage that split off 12 million years ago, the gorilla lineage that split off 8 million years ago, and then the chimpanzee lineage that split off 6 million years ago. Each went on to decrease their modal number to just four or three lumbars. Through all of this we would always have a fully upright human ancestor retaining the primitive number of five or six lumbars that may have appeared in *Morotopithecus* 21 million years ago. The oldest complete lumbar spine of a hominine—*Australopithecus africanus*—has six lumbar vertebrae.

In this model, the upright bipeds were always relatively more successful. The ultrashort-backed apes (with three or four lumbars) evolved to try to occupy niches in which quadrupedal or quadrumanual locomotion provided some ecological living space into which the upright bipeds would not relentlessly invade. The very short horizontal design has always been a marginal or losing proposition among hominoids, hemmed in as they are by the quadrupedal monkeys on one hand and the upright bipeds on the other. When we wonder why there are thousands of fossils left behind by the upright bipeds and virtually none left by the more horizontal hominoids, we have to consider the possibility that the horizontal forms were always small in number and on the verge of extinction, as their niches were encroached by monkeys and by hominoids maintaining a "less derived," primitive, hominiform body plan (that is, the upright bipeds).

Bipedal Pattern and Lumbar Configuration in the Hominiforms

Respiration and Running

The numbers alone do not really reflect the overall morphologic transformation that has occurred in the ultra-shortening of the great apes. As noted previously, in chimps and gorillas, the shift to three or four lumbar vertebrae has been accompanied by other changes that virtually obliterate all lumbar lateral bending (see figures 10-3 and 10-4). In addition, twisting of the spine is greatly limited in the lumbar spines of the great apes. These limitations are particularly important during bipedal locomotion.

Because the rib cage reaches the iliac crests in the great apes, any significant lateral bending during running will impair diaphragmatic ventilation. In fact, one of the striking differences between running in a chimpanzee as opposed to running in a human is that in the chimp, the upper body tilts towards the "stance-phase" leg, or the leg in contact with the ground (Jenkins 1972); in humans, the upper body tilts in the opposite direction toward the "swingphase" leg. The result is that the chimp's upper torso appears to swing from side to side in walking and running, while the human upper torso appears to float steadily in a static position based on the center of gravity of the body. This effect is achieved by compensatory bends and twists in our long flexible lumbar region, which almost always moves in a direction opposite from that of the pelvis. These opposite movements in the lumbar spine dampen the motions of the pelvis, thereby isolating the rib cage, arms, neck, and head, all of which remain relatively stable in humans during bipedal walking and running (Schache et al. 2002).

The anatomy of the lumbar spine is an important issue in running for several reasons. Because the rib cage is isolated from the rest of the body during running, the important result is that respiratory capacity is maintained, and can be increased even as running speed is increased. In hylobatids, during running on a branch or bipedal walking up a vine, the independent movement of the upper torso is helpful for balance. The upper body can always be moved in a direction opposite from the tilt of the pelvis. In brachiation, however, an opposite set of imperatives apply. During arm-swinging, particularly in a large ape, the pelvis and legs need to be maintained in line with the upper body as the complex inertial forces are applied to the center of gravity. This helps to maintain a more perfect pendulum. Reduction of trunk length and elimination of muscularly controlled bending in the lumbar region helps assure that respiratory function will be preserved as the speed of brachiation is maintained or increased. The pelvis will move with the rib cage rather dangling and twisting with each overhead "stride."

It certainly seems that running on the ground is a very common activity for any large-bodied hominoid. Further, it takes place in a selective arena in which there are many other ground-running predator and prey species. Although brachiation over an extended distance might seem to be a rare event, it is nonetheless a situation that can attract considerable selective pressure for an ape. Competition in this type of activity comes from other ape species in the same territory and from other individuals in the same species involved in altercations over dominance, territory, and mates.

Essentially, the maintenance of a flexible lumbar spine is an essential component of well-adapted bipedalism. The presence of five or six lumbars in

Oreopithecus, the hylobatids, and *Australopithecus* certainly leaves open the possibility that there has always been a lineage of hominiform hominoids with five or six lumbar vertebrae, and that the hominines evolved from one of these longer-backed hominoids with adequate flexibility for well-adapted bipedal locomotion.

Lumbar Stiffness and the BHBK Gait

There has been a great deal of dispute among paleoanthropologists arising from Jenkins's widely cited 1972 study of the mechanics of chimpanzee bipedalism. Using high-speed cinefluoroscopy (movie X-rays), Jenkins documented the pattern of chimpanzee pelvic movement during what has been called bent-hip, bent-knee (BHBK) bipedalism. Some have suggested that *Australopithecus afarensis* was only capable of this clumsy BHBK bipedalism and thus was still more of an arboreal creature. However, in the 30 years that have elapsed since Jenkins's publication, two things have become clear. Firstly, human-style, extended-leg bipedalism has tremendous benefits in terms of the energetic cost of locomotion. Secondly, even the modern African apes can be trained to employ an energetically efficient, extended-leg bipedal gait (Wang et al. 2003). This makes a compelling case that all of the various minor anatomical features of the pelvis, knee, and foot that have been cited as requiring BHBK bipedalism really have no such mandatory effect. The expectation that every hominoid except for the species of the genus *Homo* swayed along with laborious, ineffective, occasional bipedalism seems to be driven by the deeply held "Washburn" concept of a slow progression toward standing up, rather than by any hard evidence. The work of Robin H. Crompton (1998) and Weijie Wang, both from the Department of Anatomy at the University of Liverpool, and their colleagues has done a great deal to resolve this scientific misconception. The work by paleontologists such as Meike Kohler and Salvador Moya-Sola (1997), and Lorenzo Rook (1999) on *Oreopithecus* has also helped to clear up misconceptions about what anatomical features do or do not signify a primarily upright bipedal stance.

Owen Lovejoy, an anthropologist based at Kent State University in Ohio, has correctly pinpointed the stiff lumbar spine as the key that commits natural locomotion in African apes to quadrupedal knuckle-walking, and a preference (but not requirement) for a BHBK gait when bipedal. In a recent detailed set of review articles, Lovejoy points out that this configuration is extremely disadvantageous for bipedal walking and seems to make sense only for quadrupedal locomotion (Lovejoy 2005b, 2005a, 2007). Others have tried to explain the anatomical features that helped our ancestors shift from the "bent-knee, bent-thigh" bipedal gait used by chimpanzees into the smooth, efficient, extended-leg, bipedal gait of a human. However, Lovejoy shows that the bent limb is not a consequence of inadequate leg anatomy in the chimpanzee. Rather, it is the consequence of the inability of the chimpanzee spine to allow for adequate positioning of the torso over the stance-phase leg during bipedal walking. A flexible lumbar spine is required to make this possible. Essentially, the problem is in the spine and not in the legs.

Progress in Bipedalism Since the Human-Chimp Split

The split between the lineage of the chimpanzee and the lineage of our hominine ancestors took place around 6 million years ago. The fossil record from that time forward demonstrates a series of species that gradually become more and more similar to our modern human appearance. The individuals of the genus *Australopithecus* were more similar to the individuals of the genus *Homo* from the neck down than they were in their cranial anatomy. The australopithecine species appear to have retained and further optimized the ancient upright walking preference that dated back millions of years to the early hominiforms such as *Morotopithecus*. The succession of species of *Homo* demonstrates further progressive optimizations of bipedal walking.

A skull that is oriented on top of the cervical spine, as it is in the australopithecines, suggests that the bearer of the skull walked upright. It is this characteristic, rather any measure of increased brain size, that has been taken as the first sign of incipient humanity. If upright posture and bipedalism are primitive or ancestral traits (symplesiomorphy), then it is only a good mark of the onset of "humanity" if we extend that term back to include *Morotopithecus*, as well as a series of ancestors of the existing apes. This is the sense in which I have used the term *hominiform* in this book. The issue of how to determine the dividing line for the onset of humanity remains an interesting question.

The Pelvis and Spine in Australopithecus

The only complete lumbar spine we have from the australopithecine era is from an individual from the species *Australopithecus africanus*, designated STS-14. Because the entire lumbar spine was found together, it is possible to demonstrate that this individual actually had six lumbar vertebrae—longer than in most modern humans. This trend is certainly at odds with the evolutionary pressure leading to the shortening of the lumbar spine of the apes. The anatomy in STS-14 reflects the primitive configuration of a flexible lumbar spine that underlies our own spinal anatomy. It might be chance alone that led to the discovery of a rare australopithecine individual that happened to have six lumbars. However, it makes the great ape modal numbers of three or four lumbars even more unlikely. No orangutan, chimpanzee, or gorilla has ever been found to have six lumbar vertebrae.

The sacrum and hip bones of a now famous individual from the species *Australopithecus afarensis* were carried by Don Johanson from their ancient resting spot in modern Ethiopia to a massive vault in Cleveland, Ohio, near Case Western Reserve University. These are the remains of "Lucy" (AL-288, to be technical), and they indicate antecedents of human anatomy that may date as far back as 3.6 million years (Johanson and Taieb 1976, Wood 2002, White et al. 2006). I am grateful to Don Johanson for allowing me to examine them there in the late 1970s. Today, there are at least 21 different extinct hominine species that have been discovered. The various species are attributed to the genera of *Orrorin, Sahelanthropus, Paranthropus, Kenyanthropus, Ardipithecus, Australopithecus,* and *Homo*. The number of individual fossils now discovered that document the 6 million years of hominine evolution now reaches into the thousands. The phenomenal degree to which the past has yielded up hard evidence of our anatomical history is truly a dream come true.

In general, the modern human sacrum is much wider than the sacrum of the apes. Before the discovery of Lucy, it was believed that this was part of a mechanism for expanding the circumference of the pelvic opening in order to allow the birth of the large-headed infants of the human line. However, at the time Lucy lived, about 3.6 million years ago, the head size of an australopithecine was no different from the head size of an ape, but the sacrum was already widened. It may well be that the sacrum widened as part of the functional reconfiguration of the spine, which subsequently made it possible to have safe childbirth when larger brains evolved later on.

Evidence From the Back Muscles

I had the opportunity to do detailed dissections of the lumbar back muscles of apes that had died in the National Zoo in Washington, D.C. Their bodies had been carefully preserved for decades in storage areas of the Smithsonian Museum, and I was allowed to analyze them. Aside from gross dissection, it was possible to study the muscles for fine details such as muscle fiber length and orientation, attachment points, weight of muscle, cross-section, and potential contractile excursion, as well as a variety of other arcane details. From this data it became possible to understand how much power could be developed by the muscles, and how far they could move along any particular line of contraction.

These explorations confirmed theories of other experts that the lumbosacral musculature of the apes is much less powerful than the similar set of muscles in humans. This helps explain some of the unique features of the sacrum of *Australopithecus*, but also explains ways in which the modern human lower spine is advanced beyond the anatomy of australopithecines. Essentially, the broad sacrum of the early australopithecines provided an increase in attachment area for the muscles that reach up from the sacrum to attach along the lumbar spine. These are some of the great muscles of the *erector spinae* complex, which help carry our weight as we walk, and work to counterbalance the pull of whatever we carry in front of us, whether it is a load of building materials for a home or an infant.

Sacral tilt and expansion

To further aid in this mechanical effect, the sacrum in the australopithecines was already tilted backward to a small degree. In apes, the sacrum is parallel to the spine, but even in early australopithecines such as Lucy, the sacral tilt can be seen. This tilt progresses further in modern humans, but in both cases this helps place the origin of the spine muscles well behind the position of the lumbar vertebrae. This design also lends extra leverage to the muscular power of the lumbar spine.

Reorientation of the iliac crest

In addition to the broadening of the sacrum, the power of the erector spinae muscles in humans is increased by one final modification. Many of these muscles arise on the upper part of the pelvic hipbone or *iliac crest*. In apes, the blades of the iliac crest are essentially flat and parallel to each other; however, in modern humans, these blades are turned backward, and project out behind the lumbar spine as far as two or three inches. This rearrangement brings the main origin of the back muscles out to a position well behind the curves of the lumbar spine. As with the widening and tilting of the sacrum, the effect of the reshaping of the iliac crests is to greatly increase the leverage of the spine muscles, thus helping to hold the body erect when carrying weight in front.

The broadening of the sacrum, the reorientation of the sacrum so it projects out posterior to the lumbar spine, and the remodeling of the iliac crests (Filler 2002b) all work together to help improve upon the design of the modern human. In fact, as with the opposable thumb and its role in promoting the evolution of tool use, the modification of the lumbosacral anatomy has an important role in setting the stage for the most remarkable aspects of human evolution. A powerful lumbar spine allows the individual to carry heavy objects in front while walking and lifting. This helps open up the universe of what the hands and arms can do.

Evidence From the Feet: Heel-Strike and First Toe

Presence of an opposable big toe does not necessarily prove that terrestrial bipedalism was slow or inefficient. It now seems clear that hominines such as *Australopithecus afarensis*, who certainly were primarily bipedal, retained an opposable big toe (Harcourt-Smith and Aiello 2004). Because of the use of heel-strike footfalls, rather than stepping with the mid-foot impacting first, any negative effect of a separated big toe upon the efficiency of walking is minimized. Although heel-strike walking is an influence on the bones of the talus, calcaneus, and other parts of the foot, there is a possibility that running has played a role in these architectural features, and may have done so as well in creatures such as the common ancestor of the chimps and the hominines.

In human running, the front part of the foot hits the ground first, so that we can employ the elastic recoil of the landing to help power the next takeoff in the stride. The major energy storage and release event involves the calf muscles and the Achilles tendon. When a running human lands on the mid-foot, the entire weight of the body—carrying a momentum up to three times greater than during walking—crashes downward over the middle/anterior part of the calcaneus. The body pushes down, but the Achilles tendon pulls up. Therefore, the massive calcaneus is needed in order to assure that this bone is not snapped in two at the first running step.

The role of distance running

A newer perspective on this and a variety of other anatomical features of human bipedalism has shifted the focus from walking to distance running (Bramble and Lieberman 2004). This is interesting for two reasons. Firstly, although many primates can walk bipedally and can run short distances bipedally, only humans are effective long-distance bipedal runners. Secondly, it is easy to see that life events that require running, such as chasing down game and escaping predators and assailants from your own species, are particularly likely to affect survival. Therefore, they will be subject to strong adaptive pressure under natural selection.

Long-distance running is one of the great skills of humans when compared to other species. Although a horse is far faster in a sprint or a short gallop, a well-trained human can outrun a horse if the distance to be traveled is greater than 15 kilometers, or about 10 miles (Bramble 2004). Based on the anatomy of fossils, the current thinking is that the capability for extended running appears to be a relatively late development. Many features related to this ability are seen in *Homo habilis* at 2.5 million years ago, and virtually all of them are seen in *Homo erectus* at 1.8 million years ago, but there is little evidence for running specializations in *Australopithecus*. However, this view is still open for interpretation.

The Theoretical Impact of Persistent Hominiform Bipedalism

Continuous Bipedalism vs. Multiple Origins of Bipedalism

Lovejoy, as does Pilbeam, supports a progressive "relengthening" of the ultrashort spine of an ancestral knuckle-walker in the eventual evolution *of Australopithecus* and *Homo* to achieve efficient bipedal walking (Lovejoy 2005a). However, there is no evidence in Lovejoy's model to prove that an ultrashort-backed condition requiring a bent-leg gait existed in the last common ancestor of the great apes and humans. There just isn't any basis in the fossil record for this position. All the fossils ever found from the Miocene onward just simply fail to show this. They all show the hominiform version of a long, flexible spine with five or six lumbar vertebrae.

By contrast, Preuschoft, an anatomist from the Ruhr University in Germany, makes a very different point in his extensive review published in 2004. He draws attention to the frequent and effective use of bipedalism by the orangutan. *Oreopithecus* (the fossil ape with evidence of a long, flexible spine and habitual bipedalism) and the lineage of the orangutan both split off from the lineage of African apes and humans around 12 to 15 million years ago. These two species provide further evidence in favor of a major role for bipedalism in the last common ancestor of the man, the great apes, and *Oreopithecus*. There is no obvious fault to this logic, and it certainly has the direct support of the fossil evidence. The question of an upright bipedal ancestor then comes down to predicting what happens in the lineage between the divergence of the orangutan and the divergence of the gorilla 3 to 4 million years later. Either the bipedalism and upright posture with a longer spine are maintained, or they are briefly eliminated while the gorilla branches off, and then suddenly reestablished again immediately after the chimpanzee branches off. This may have happened, but because there is absolutely no fossil evidence of any kind for this "up, up, up, up, down, up" sequence, it is hard to accept that it is the only reasonable theory.

No convincing argument has been made for a unique ecological alteration or other event that could accomplish this odd fluctuation. This idea of a knuckle-walking intermediate in human evolution requires a peculiar sequence of events. There would be a long flexible spine for 21 million years, except for a period of 3 million years between the branching of the gorilla and the branching of the chimpanzee. In that comparatively brief period, the spine would be shortened and a primarily upright posture lost along with primary bipedalism in favor of knuckle-walking. Then, both would reappear, fairly suddenly after the divergence of the chimpanzee lineage, for the birth of the hominine lineage—again, for no obvious compelling reason. The new bipedalism is treated as effectively unrelated to the old "pre-knuckle-walking" bipedalism. Parsimony suggests that the proposed brief disappearance of primary upright bipedalism, which was then replaced with a new and different secondary bipedalism, may never have actually taken place.

At the time the hylobatids branched off 17 million years ago, we have a lineage resulting in an upright biped with a flexible lumbar spine of five or six lumbar vertebrae. When the more detailed fossil record of *Australopithecus* commences at 2.5 million years ago, we have an upright biped with a flexible lumbar spine of five or six lumbar vertebrae (Robertson 1972). We also have evidence in Europe for *Oreopithecus* at 12 million years ago, which apparently was an upright biped with a flexible lumbar spine of five or six lumbar vertebrae (Rook 1999). What is missing here is any solid reason to deny the possibility that all along the way, from *Morotopithecus bishopi* to *Homo sapiens*, there has always been a hominoid in our own ancestral lineage that was an upright biped with a flexible lumbar spine of five or six lumbar vertebrae. In this view, the human lumbar spine represents the persistent primitive condition with a most common occurrence of five lumbar vertebrae. The lumbar vertebra of *Morotopithecus* is undoubtedly the vertebra of an upright biped.

Ecological vs. Mutationist Explanations for Bipedalism

In the 1999 edition of his textbook *Primate Adaptation and Evolution*, John Fleagle, a leading primatologist and anthropologist from the State University of New York at Stonybrook and winner of the MacArthur "genius" fellowship, has summarized and discussed 10 different theories on why and how the emergence of bipedalism could have resulted from various adaptive pressures. These include carrying tools, weapon use, carrying infants and food, traveling between trees, feeding from fruit in tall bushes, feeding on seeds from tall grasses, the male carrying food for the family, reducing exposure to the sun (only the top of the head gets the full heat), looking over tall grasses, and aquatic life. He dismisses all of them as unsatisfactory in various ways and offers no further suggestions.

One theory that Fleagle does not consider is the possibility investigated in this book—namely, that the common ancestor of chimps and humans was already a bipedal ape. This is, at its fundamental core, a mutationist rather than an adaptationist view (Stoltzfus 2006). The potential for a sudden arrival of the anatomical conditions to support upright bipedalism in the course of a single generation appears to be revealed in the anatomy of *Morotopithecus*. The driving force may not have been ecology, environment, and selection, but rather the emergence of a new hominiform body plan, one that was subsequently adjusted and optimized by the pressures of natural selection, but not initiated by classical Darwinian processes. This is comparable to the events that led to the origin of the vertebrate body plan. Modern genetics has borne out the vision of Étienne Geoffroy Saint-Hilaire on this point. Rather than a gradual progression from invertebrates to vertebrates across tens or hundreds of millions of years, the origin appears to have taken place in a single generation with the flip of the embryonic body axis. The body axis did not gradually rotate 180 degrees in a series of 180 steps of one degree each across millions of years, as a result of rotatory natural selection pressures acting on the relative reproductive success advantages of variants. It flipped suddenly, in one individual and one generation. It did so because of a mutation, and as a result, a new body plan that defines our phylum was abruptly established as a basis for further diversification and optimization by natural selection over the subsequent hundreds of millions of years.

The evidence in the hominoid fossil record, taken together with our recent understanding of the effects of mutations in the *Pax1* and *Pax9* morphogenetic genes, suggests a similar sudden body plan transformation leading to the origin of the hominiform body plan. Subsequently, Darwinian adaptation plays a role in refining the architecture and generating various versions of the principal body plan. However, a modern modular view of evolutionary change is essential to understanding human origins.

Parallel Evolution of Quadrupedal Recidivism

Why did the three great ape lineages—orangutan, gorilla, and chimpanzee—abandon full-time, upright, orthograde, bipedal locomotion? Why do gibbons, siamangs, and humans persist in their strict, upright, orthograde bipedal locomotion? Is there any benefit to a flexible mix of bipedal and quadrupedal locomotor options? The parallel reappearance of quadrupedal walking in three of our close relatives is a puzzle to many who think about human origins. They want to know the adaptive scenario—for example, how this is impacted by forests, vegetation, tree canopies, grass types, the spread of the savannah, earlier global warming and cooling, and so on. However, all adaptive scenarios for the evolutionary drive behind human bipedalism or great ape quadrupedalism are post hoc, rationalized hypotheses that are more or less immune to proof or verification. They also tend to be disengaged from any fixed systematic or temporal constraints. For instance, virtually all of the ecological arguments for the cause of the supposed emergence of bipedalism in *Australopithecus* seem to apply equally well to a common chimp-human ancestor 1 million years earlier. Most also apply to virtually any hominiform from *Morotopithecus* onward because of the powerful anatomic predisposition to upright posture present in all of the hominiforms, due to their shared modification of the lumbar spine.

Adaptive pressures against bipedalism

Given all the pressures for progress in bipedal form and capability, it is interesting to consider why the African apes evolved backward toward a quadrupedal style of locomotion. In one scenario, the spine of the apes evolved above and beyond the more primitive lumbar anatomy of the proto-humans, broadening the array of accessible niches and allowing the apes to return to a desirable life in the trees. An alternate view is that the chimp and gorilla ancestors were driven into the less-desirable forest to avoid and escape the proto-humans. Only the best walkers, runners, and carriers could compete for control of the more lucrative ground environment. The ancestors of the apes retreated into the forest, as they were not optimally designed for such a life at the outset. To survive and be successful in the trees with the lumbar spine of a human, additional changes were needed. The modifications and extreme reduction of the lumbar spine that we now see in the great apes are ways of modifying a poorly designed body to function well in the only habitat not overrun by the early proto-humans.

Advantages of a quadrupedal capability

The evidence from across the board in the mammals is that quadrupedalism is a great way to move. In fact, it seems quite clear that in the Late Miocene and Pliocene of Africa, the quadrupedal catarrhine primates (Old World monkeys such as the baboon) enjoyed tremendous success relative to the orthograde apes. As the number of species and individuals of the monkey group exploded, the apes and human ancestors seem to have just barely persisted by comparison. The point is that we need to keep in mind the possibility that upright bipedalism can be a bit of a disadvantage. A monkey can outrun a human, and a chimpanzee runs faster on all fours than it does bipedally. One simple benefit of quadrupedal progression on the ground appears to be increased speed, and this is particularly true if the bipedal gait is inefficient. In this scenario, upright bipedalism/suspensory posture is pushed into the hominoid lineage in the Miocene by a morphogenetic change that is basically a birth defect. It persists because brachiation and suspension become possible and open a useful niche for feeding in trees. However, in many situations it can also be an anatomical trap, because orthograde primates seem to be poor competitors against pronograde primates. To successfully survive as an orthograde primate in a pronograde primate world, the lineages of orangutans, chimps, and gorillas have each produced very significant alterations that add to or modify the *Morotopithecus* lumbar architecture to allow for diagonograde progression on the ground. This has apparent benefits in stability, gait flexibility, and speed.

The human lineage can be seen as a group of hominoids burdened with a spine that requires upright posture, and whose lineage failed to evolve an architecture to allow them to achieve diagonograde posture. This view is premised upon viewing the ancestral condition of upright bipedalism as a *disadvantage* imposed upon the lineage by the Moroto *Hox* and *Pax* changes in morphogenetic DNA. To survive in competition with the monkeys and the diagonograde, large-bodied apes, the hominines evolved compensating advantages such as more complex hand use, progressive increase in brain size, and ultimately, symbolic language (Deacon 1997). In addition, hominine adaptive evolution made bipedalism more viable. Through a series of changes in hip structure, leg length, foot construction, knee posture, and lumbar modifications, humans eventually were able to employ an improved bipedalism that was highly effective in competition with quadrupeds.

Mutation and Adaptation in Hominoid Locomotor Patterns

All of this demonstrates an interesting feature of evolution above the level of the species. We have an evolutionary innovation introduced by mutation that produces a mix of advantages and disadvantages. Lineages emerge that fine-tune and optimize in various directions. The new spine of Moroto prevents comfortable pronograde quadrupedal walking. Some lineages develop "workarounds" that modify the spine a second time to allow for diagonal postures that recover many of the advantages of quadrupedalism. Other lineages proceed to optimize the bipedalism in various ways and with various levels of effectiveness. For example, *Oreopithecus* will have developed a different set of optimizations of bipedalism than did the hominines.

From a perspective of strict Darwinian gradualism, there is an assumption that there will be an infinite supply of variation in every direction for every character. This will allow gradual progression in any advantageous direction. In the mutationist view highlighted in this book, evolution can only proceed along new lines—such as new body plans—if mutations are produced that create innovations (Stoltzfus 2006). Earlier in time, "horizontal" events created such new designs. The ultimate example is the day when an early, one-celled organism engulfed a cyanobacterium (see Chapter 7). Genetic analysis shows that all the plants and algae on the planet trace their existence to that sudden event. Photosynthesis did not arrive in the eukaryotic cells by gradual Darwinian evolution acting upon variation. Although natural selection has subsequently produced a marvelous variety of plants in the world, and although it produced the biochemistry of the cyanobacterium, it did not gradually accomplish the constructive steps to lead to photosynthesis in the "vertical" lineage of the plants. The point is that the key events in the History of Life are not all strictly Darwinian.

Selection at multiple levels as proposed by Gould (2002) accommodates a wide variety of heritable or "vertical" transformations in a hierarchical or modular selection theory. In general, my argument is that major changes such as the flip of the dorso-ventral axis that produced the vertebrates, or the dorso-ventral crossing of the spinal cord axis that produced the post-Moroto hominoids, involve major transformations that can take place suddenly. Such changes are capable of being captured in chromosomal speciation events to launch new types of animals in new lineages. Darwinian gradual selection and adaptation then takes on the more drawn-out task of fine-tuning and optimizing the various features of the new type of organism.

Selection at the level of the High Order Modules—dorsal versus ventral, limbs versus somites—certainly may have played an initiating role in hominiform origins. There are plenty of compelling reasons for the propagation and fixation of the genes involved in hominoid upright posture, genes that have their origin in more commonly understood type of selection between species members and between competing species. However, because of the type of change that did occur, and because of the impressive fixation (lack of change over time) over the change in homology of the LTP, it is possible that an event dominated by morphogenetic mechanisms played a very important role. Indeed, this is what appears to have happened with the Moroto transformation. A new type of primate was produced that had a lumbar spine very well-suited for upright postures. However, this spine did not allow for comfortable pronograde postures during locomotion, or the improved speed and stability of quadrupedal running.

The lineages of the orangutan, the gorilla, and the chimpanzee broke out of this limitation with their various quite independent modifications of the Moroto lumbar architecture, which has allowed them to revisit the proven success of quadrupedal progression. The hominine lineage never accomplished this. Instead, it achieved other types of features that have allowed their survival and ultimate grand success, even with the requirement of obligatory bipedal progression on the ground. Darwinian adaptation has clearly converted bipedalism from a relative burden to a wonderful advantage. This is demonstrated by the numerical and geographical success of our species relative to other primates. Optimizations of bipedal progression, such as our ability to accomplish long-distance running (Bramble 2004), have also improved the viability of bipedalism. In addition, at some point a decisive moment was reached in which hand and arm use for carrying and throwing during bipedal locomotion became essential. In this setting, any hominine that evolved a way to locomote quadrupedally would be at a disadvantage relative to the other hominines. Thus, bipedalism became a totally fixed trait in subsequent hominine species.

Humanity's origin seems to have stemmed from a morphogenetic mutation that generated the hominiform body plan. This event generated a species of hominoid that suddenly displayed most of the numerous major anatomical differences between the hominoids and the monkeys. When a sudden change such as this is proposed, it is important to provide evidence for the time frame as well as for the genetic mechanism. The sudden origin of the vertebrates is convincing because the genetic event that caused the embryonic body axis to flip 180 degrees can be detected, and because it is nonsensical to argue that the body slowly flipped a few degrees at a time over millions of years. For the hominiform origin, we are bracketed by the hominoid-cercopithecoid (ape-monkey) split at 23 million years, and the existence of the fully transformed *Morotopithecus* at 21.6 million years. There is also evidence that *Proconsulid* dental apes (hominoids by their dental pattern that do not share our modified body form) emerged and diversified before *Morotopithecus* appeared. Discovery of intermediates between the *Proconsulid* body plan and the hominiformid body plan may yet be discovered. However, given the possibility that most of the major body plan changes were connected to a related set of morphogenetic processes moderated by a single gene (*Pax1*), it is certainly possible that there never were any intermediates.

After the sudden origin, upright bipedal walking with many similarities to our own, and an anatomy from the neck down that was in many ways similar to the human form, appeared early and persisted. Why are there so many upright bipeds in the history of the hominiform hominoids? One well-supported answer is that they share a common ancestor that had the face and brain of an ancient ape, but the upright bipedal form that was fundamentally human in its aspect. Recently, looking forward from our new vantage point with the completely decoded genome of humans and chimpanzees now in hand (Mikkelson 2005), Sean Carroll described the new challenge of deciphering human evolution (Carroll 2003). Although certain behaviors and attributes, such as symbolic thought, language, and higher consciousness, are key features of humanity (Deacon 1997, 2000), they are not all that promising for testable reduction to individual gene changes. The situation is easiest when we have simple biochemical differences from the chimpanzee, such as differences in hemoglobin, for example. However, these changes rarely inspire a perception of essential humanity.

Although many people still have a difficult time accepting that an ape-like creature could have been transmuted into a human, there are few who would seriously contend that there is any fundamental major qualitative difference in biochemistry, biology, or physiology between a human and a closely related primate (see Figure 10-6). Even with regard to our amazing brain, brilliant neurobiologists such as Terrence Deacon have searched with all of the most advanced techniques of modern neuroscience, and there is no part of the human brain that does not exist in the chimpanzee (Deacon 1992, 1997, 2000, 2008). Our most accomplished molecular geneticists have explored the genomes and found no differences any greater than those that exist for any other pair of related species separated in divergence times by 5 or 6 million years (Ruvolo 1994, 2004).

Irven DeVore, in his 1962 Ph.D. thesis, showed that baboon societies demonstrate a surprisingly extensive array of complex social behaviors previously thought of as the sole prerogative of humans. As his work has been repeated by others, much of what had previously been considered as strictly human behavior proved to be not all that different from the social complexity of apes or monkeys. Detailed observations of the natural behavior of chimpanzees, as well as their capabilities for learning to communicate with hand signs, facial gestures, art, and simple symbols, have placed even more limits on what can be claimed as unique for humans.

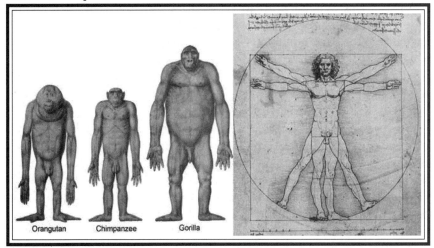

Figure 10-6: The nature of humanity. The relative body proportions of the four great hominiforms. Drawings of the apes by Schultz (*left*) reflect the foot positioning used by DaVinci when he captured the measure of man (*right*), as described by the roman architect Vitruvius.

If all that is left is our relentless progression in technology and our complex symbolic language (Deacon 1997, 2008), then we have a different sort of problem with our definition of humanity. There doesn't seem to be much evidence of technological advance or any detectable sign of our language complexity until about 35,000 years ago. If we require these features for a definition of humanity, then not only do we have to exclude all the australopithecines, but also *Homo habilis* and *Homo erectus*. In fact, even within our own species of *Homo sapiens*, we must exclude sub-species such as *Homo sapiens neandertalensis* (the neanderthals), and we more or less have to exclude many generations of our own sub-species of *Homo sapiens sapiens*, as they showed little sign of difference from the neanderthals from their first appearance 100,000 years ago until art begins to appear 65,000 years later.

No one really advocates classifying a neanderthal as an ape. However, upon what facts do we base our assertion that chimpanzees do not meet the qualifications for being human? Ever since the impact of DeVore's 1962 study had sunk in, we have had a safe answer that keeps the chimpanzees and other apes out of the sacred temple of humanity—namely, our upright posture and habitual bipedal walking. Who first proposed this as the critical innovation of humanity? Well, it was Irven DeVore's thesis advisor, Sherwood Washburn. By advocating a stooped quadrupedal collection of apes as distinct from the nobly upright human, Washburn had defused the potential crisis—for a while, that is. Now however, the problem with Washburn's solution is being brought into sharp focus. When I wrote to Washburn to tell him what I had found, he alone knew best what would follow. He acknowledged that the relevance of bipedality in the early hominoids in general may have been underemphasized in his work.

What does it all mean? What we have mistaken for humanity is, indeed, just language and technology. Humanity itself had its origin more than 21 million years ago. The origin was not a matter of emotion, of knowledge, or of intellect; rather, it was a matter of morphogenetic change. And so we have a new possibility: a first human child, born to an ape parent, awkward and erect, and always at a loss to keep up with its quadrupedal proconsulid siblings, but nonetheless the harbinger of a new and remarkable species. The dawn of the great hominiform lineage was almost certainly then captured by a chromosomal speciation event. This true first child, the founder of humanity, was the product of a nearly instantaneous, but nonetheless truly awesome event in the History of Life on this planet—namely, a massively pleiotropic mutation in the *Pax* gene system in a gamete of that one ape mother during the Early Miocene. The first upright ape was also the first human. In the millions of years that followed, new species branched off and abandoned their upright posture to descend to what we now call "ape." However, from that moment of human genesis, there has always been an upright ancestral lineage of species, keeping the gleam of promise alive until it blossomed fully into the literate technological and inventive creatures that we are, able to reflect in wonder and understanding at the path by which we came to be.

Figure Credits

Figure 1-1: From Owen 1848, p. 81.

Figure 1-2: From Owen 1848, Plate II.

Figure 1-3: Photograph by F. Rudolf Turner, Department of Biology, Indiana University, Bloomington.

Figure 2-1: From Goethe 1790.

Figure 2-2: From Goethe 1784 and 1817, Plate I and Plate IX, original drawings by Waitz.

Figure 3-1: From France 1821.

Figure 3-2A: From France 1821.

Figure 3-2B/D: From Budge 1911 (1973 Dover Edition).

Figure 3-2C: Photo courtesy of John Bodsworth, Egyptarchive.

Figure 3-3A: From the Virtual Egyptian Museum, item number MET.MM.00122.04 (*www.virtualegyptianmuseum.org*).

Figure 3-3B: From Andrews 2004. Copyright by Trustees of the British Museum.

Figure 3-3C/D: From France 1821.

Figure 4-1: Drawing by Aaron G. Filler.

Figure 4-2A/B: Reproduced with permission from the Office of Science Policy, the National Institutes of Health, the National Center for Biotechnology Information, and the National Library of Medicine, *www.ncbi.nlm.nih.gov/ About/primer/genetics_cell.html*.

Figure 5-1: Reprinted from Kuratani 2005a, pp.1–19. Copyright 2005 by the Zoological Society of Japan.

Figure 5-2A: From Bolton 1889, p. 49.

Figure 5-2B: Painting by Josef Stieler 1828.

Figure 5-2C: Drawing by Ambroise Tardieu, Smithsonian Institution Library.

Figure 5-2D: Photograph by Maull and Polyblank. Copyright by The Royal Society, London.

Figure 5-2E: Photograph by Maull and Polyblank. From Pearson 1914.

Figure 5-2F: Photograph courtesy of the National Library of Medicine, National Institute of Health.

Figure 5-2G: Photograph courtesy of the University of Cambridge.

Figure 5-2H: Photograph by Gerd Hennig, GNU Free Documentation License.

Figure 5-2I: Photograph from the Archives of the Ernst Mayr Library of the Museum of Comparative Zoology, Harvard University.

Figure 5-3: Redrawn and modified after Hennig 1966, pp. 89 and 148.

Figure 6-1: From Hobbes 1661.

Figure 6-2: Reprinted from Penny and Phillips 2004, pp.516–522. Copyright 2004 by Elsevier. Used with permission.

Figure 6-3A: From Aaron Filler and Andrew Lever, U.S. Patent 5,554,498.

Figure 6-3B: Reprinted from Alberts et al 2002. Copyright 2002 by Garland Science/Taylor and Francis LLC. Reproduced with permission.

Figure 6-3C: From Filler and Lever 2007.

Figure 6-3D: Redrawn and modified after Harrison et al. 1998.

Figure 7-1A: Redrawn and modified protostome and deuterostome developmental sequence after Kardong 2005.

Figure 7-1B: Redrawn and modified figures of *Porifera* and *Nematoda* after Hickman 1973.

Figure 7-1B: Redrawn and modified figures of *Cnidaria* and *Platyhelminthes* after Hickman 1955.

Figure 7-1B: Redrawn and modified cladogram after Kardong 2005.

Figure 7-1B: Redrawn and modified figure of squid after Drew 1911.

Figure 7-1B: Redrawn and modified animal figures of *Arthropoda*, *Echinodermata*, *Hemichordata*, and *Chordata* after Liem et al. 2001.

Figure 7-2A: Redrawn and modified after Elder 1980.

Figure 7-2B: Redrawn and modified after Sleigh and Barlow 1980.

Figure 7-2C: Redrawn and modified after Dickinson et al. 2000.

Figure 7-3: Redrawn and modified after Ikuta 2005.

Figure 7-4ABC: From Kardong 2005, p. 297. Copyright 2005 by The McGraw-Hill Companies, Inc.

Figure 7-4DEF: Redrawn and modified after Liem 2001.

Figure 7-5A: From Raven et al. 2005, p. 399, fig. 19.20. Copyright 2005 by The McGraw-Hill Comapnies, Inc.

Figure 7-5B: From Kuratani 2005a. Copyright 2005 by The Zoological Society of Japan.

Figure 8-1A: Photo by Aaron Filler.

Figure 8-1BD: From Kardong 2005, p. 345, fig. 9.31; p. 390, fig. 10-28. Copyright 2005 by The McGraw-Hill Companies, Inc.

Figure 8-1C: Redrawn after Benton 2005.

Figure 8-1C: C1-1, C2-1, C1-2 from Jenkins 1971, p.169, fig. 48-II; p.175, fig. 51A; frontispiece. Courtesy of the Yale Peabody Museum of Natural History.

Figure 8-1C: C2-2 from Romer and Lewis 1973. Reprinted with permission from the Harvard University Museum of Comparative Zoology.

Figure 8-1C: C-3 from Jenkins 1970. Reprinted with permission from the Harvard University Museum of Comparative Zoology.

Figure 8-2ABD: From Hoffstetter and Gasc 1969; p.231, fig 21; p.245, fig. 36; p. 249, fig. 40. Reprinted with permission from Carl Gans and with permission from Jean-Pierre Gasc.

Figure 8-2C: Redrawn and modifed after Romer 1956.

Figure 8-2D: Modifed after Wettstein 1937. Used by permission of Walter de Gruyter Gmbh and Co., KG.

Figure 8-2E: Redrawn and modified after drawing in personal communication from C. Richard Taylor.

Figure 8-3AB: From Dean 1895.

Figure 8-3C: Redrawn and modified after Schmidt-Nielsen 1979; p. 29, fig. 2.3.

Figure 8-3D: Redrawn and modified after Randall 1981; p. 102, fig. 4.16.

Figure 8-3E: Redrawn and modified after Alexander 1981; p. 368, fig. 11-15.

Figure 8-3F: Redrawn and modified after Gambaryan 1974; p. 50, fig. 44. Also "Beg Mlekopitayushchikh—Prispobitel'nye." Izdatel'stvo Nauka, Leningradskoe Otdelenie 1972.

Figure 8-4AB: From Irschick and Jayne 1999; p. 1055, fig. 5. Reproduced with permission from the Company of Biologists, Ltd.

Figure 8-4CD: From Heilmann 1927.

Figure 8-4E: From McLoughlin 1979, pp. 7, 50, 51. Copyright 1979 by John C. McLoughlin. Reprinted with permission from Viking Penguin, a division of Penguin Group (U.S.A.), Inc.

Figure 8-4F: From Janensch 1950. Reprinted with permission from E. Schweizerbart'sche Verlagsbuchhandlung (Nägele u. Obermiller), Science Publisher, *www.schweizerbart.de/j/palaeontographica/E-palaeontographica.html.*

Figure 8-4G: Redrawn and modified after Bakker 1993, p. 155.

Figure 8-5A1: From Williston 1925.

Figure 8-5A2: From Hoffstetter and Gasc 1969; p. 230, fig. 20. Reprinted with permission from Carl Gans and with permission from Jean-Pierre Gasc.

Figure 8-5B: From Jenkins 1971; p. 51, fig. 12 (foldout). Courtesy of the Yale Peabody Museum of Natural History.

Figure 8-5CD: From Filler 1986; fig. 102.

Figure 8-6ABC: From Filler 1986; fig. 14, fig. 3, fig. 9.

Figure 9-1: From Filler 2007a.

Figure 9-2: From Filler 2007a.

Figure 9-3: Photographs by Nina Leen.

Figure 9-4: From Vereecke et al. 2006; p. 552–567, fig. 3, fig. 5a; Copyright 2006 by Elsevier. Reprinted with permission.

Figure 9-5: From Filler 2007a.

Figure 9-6A: From Filler 2007a.

Figure 9-6B: From Owen 1857, Plate 35.

Figure 10-1A: Redrawn and modified after Simons 1972; p. 55, fig. 12.

Figure 10-1B: Redrawn and modified after drawing by Jack Kunz, from Eimerl and DeVore 1968; p. 115.

Figure 10-1C: Redrawn and modified after Preuschoft 2004; p. 364, fig. 1.

Figure 10-2A: From Filler 1986; fig. 73, fig. 74a.

Figure 10-2B: From Owen 1857, Plate 35.

Figure 10-3A: From Filler 1979.

Figure 10-3BCD: Photograph by Aaron Filler.

Figure 10-4: Redrawn and modified after Filler 1993.

Figure 10-4: Hominoid spines from Filler 1986.

Figure 10-4: Redrawn modified drawings of animal figures of gibbon, gorilla, macaque, and lemur after Napier and Napier 1985.

Figure 10-5A: Photograph by Aaron Filler.

Figure 10-5B: Photograph by Aaron Filler, reconstructed from original medical image from Filler 1996.

Figure 10-5BD: Diagram from magnetic resonance image from Filler 2004. Copyright 2004 by Elsevier. Reprinted with permission.

Figure 10-5C: Modified after Filler 1993, pp.13–20. Reprinted with permission from Thieme.

Figure 10-6A: From Schultz 1933, 1969. Reprinted with permission from Orion and with permission from the Anthropological Institute of the University of Zurich.

Figure 10-6B: From DaVinci 1492.

Bibliography

(For additional reading, please see the suggested reading lists, organized by chapter, on *www.uprightape.net*.)

Ahlberg, P. E., J. A. Clack, and H. Blom. "The axial skeleton of the Devonian tetrapod *Ichthyostega*." *Nature* 437 (2005): 137–40.

Alberts, B. *Molecular Biology of the Cell*. New York: Garland Science, 2002.

Alexander, R. M. *The Chordates*. Cambridge, U.K.; New York: Cambridge University Press, 1981.

Allbrook, D. B., W. W. "New fossil hominoid material from Uganda." *Nature* 197 (1963): 1187–1190.

Andrews, C., and British Museum. *Egyptian Mummies*. Cambridge, Mass.: Harvard University Press, 2004.

Appel, T. A. *Cuvier-Geoffroy Debate: French Biology in the Decades Before Darwin*. New York: Oxford University Press, 1987.

Arthur, W. "The emerging conceptual framework of evolutionary developmental biology." *Nature* 415 (2002): 757–64.

Arthur, W., and A. D. Chipman. "How does arthropod segment number evolve? Some clues from centipedes." *Evolution and Development* 7 (2005): 600–7.

Bakker, R. T. *The Dinosaur Heresies: New Theories Unlocking the Mystery of the Dinosaurs and Their Extinction*. New York: Kensington, 1993.

Balling, R., U. Deutsch, and P. Gruss. "*Undulated*, a mutation affecting the development of the mouse skeleton, has a point mutation in the paired box of *Pax 1*." *Cell* 55 (1988): 531–5.

Bateson, P. "William Bateson: a biologist ahead of his time." *Journal of Genetics* 81 (2002): 49–58.

Bateson, W. *Materials for the Study of Variation Treated with Especial Regard to Discontinuity in the Origin of Species*. London; New York: Macmillan, 1894.

Bateson, W. and G. Mendel. *Mendel's Principles of Heredity; A Defence*. Cambridge, U.K.: Cambridge University Press, 1902.

Beaster-Jones, L., A. C. Horton, J. J. Gibson-Brown, et al. "The *amphioxus* T-box gene, *AmphiTbx*15/18/22, illuminates the origins of chordate segmentation." *Evolution and Development* 8 (2006): 119–29.

Begun, D. R. "Planet of the apes." Scientific American 289 (2003): 74–83.

———. "Anthropology. The earliest hominins—is less more?" *Science* 303 (2004): 1478–80.

Benton, M. J. *Vertebrate Palaeontology*. Malden, Mass.: Blackwell Science, 2005.

Bishop, W. W. "Miocene Mammalia from the Napak Volcanics, Karamoja, Uganda." *Nature* 182 (1958): 1480–1482.

———. "More fossil primates and other Miocene mammals from North-East Uganda." *Nature* 203 (1964): 1327–1331.

———. "Tertiary mammalian faunas and sediments in Karamoja and Kavirondo, East Africa." *Nature* 196 (1962): 1283–1287.

Bolton, S. K. *Famous Men of Science*. New York: T. Y. Crowell & Co., 1889.

Bramble, D.M., and D.R. Carrier. "Running and breathing in mammals." *Science* 219 (1983): 251–6.

Bramble, D.M., and F.A. Jenkins Jr. "Mammalian locomotor-respiratory integration: implications for daiphragmatic and pulmonary design." Science 262 (1993): 235–40.

Bramble, D. M., and D. E. Lieberman. "Endurance running and the evolution of *Homo*." *Nature* 432 (2004): 345–52.

Bryce, T.H. "Certain Points in the Anatomy and Mechanism of the Wrist-Joint...." *Journal of Anatomy and Physiology* 31 (1896): 59–79.

Budge, E. A. W. *Osiris and the Egyptian Resurrection*. New York: G. P. Putnam's Sons, 1911.

———. *The Book of the Dead. The Papyrus of Ani in the British Museum*. London: Longmans & Co., 1895.

Campbell, J. *The Hero With a Thousand Faces*. New York: Pantheon Books, 1949.

Carroll, S. B. "Genetics and the making of *Homo sapiens*." *Nature* 422 (2003): 849–57.

Carroll, S. B., J. K. Grenier, and S. D. Weatherbee. "From DNA to Diversity: Molecular Genetics and the Evolution of Animal Design." Malden, Mass.: Blackwell, 2005.

Cavalier-Smith, T. "A revised six-kingdom system of life." *Biological Reviews of the Cambridge Philosophical Society* 73 (1998): 203–66.

Chugg, A. "The sarcophagus of Alexander the Great?" *Greece & Rome* 49 (2002): 8–26.

Corsetti, F. A., S. M. Awramik, and D. Pierce. "A complex microbiota from snowball Earth times: microfossils from the Neoproterozoic Kingston Peak Formation, Death Valley, U.S.A." *Proceedings of the United States National Academy of Science* 100 (2003): 4399–4404.

Crompton, R. H., L. Yu, W. Weijie, et al. "The mechanical effectiveness of erect and 'bent-hip, bent-knee' bipedal walking in *Australopithecus afarensis.*" *Journal of Human Evolution* 35 (1998): 55–74.

Curran, B. A. "The Renaissance Afterlife of Ancient Egypt (1400-1650)." In *The Wisdom of Egypt: Changing Visions Through the Ages* by P. J. Ucko and T. Champion. London: U.C.L. Press, Cavendish Publishing Ltd., 2003.

Dainton, M. "Palaeoanthropology: Did our ancestors knuckle-walk?" *Nature* 410 (2001): 324–6.

Dainton, M., and G. A. Macho. "Did knuckle-walking evolve twice?" *Journal of Human Evolution* 36 (1999): 171–94.

Damen, W. G., R. Janssen, and N. M. Prpic. "Pair rule gene orthologs in spider segmentation." *Evolution and Development* 7 (2005): 618–28.

Darwin, C. *On the Origin of Species by Means of Natural Selection; or, The Preservation of Favoured Races in the Struggle for Life.* London: J. Murray, 1859.

———. *The Origin of Species by Means of Natural Selection; or, The Preservation of Favoured Races in the Struggle for Life, 6th Ed.* London: J. Murray, 1878.

DaVinci. *Canon of Proportions.* Venice: Academy of Fine Arts, 1492.

Dawkins, R. *The Blind Watchmaker: Why the Evidence of Evolution Reveals a Universe Without Design.* New York: W.W. Norton, 1996.

———. *The Selfish Gene.* New York: Oxford University Press, 1976.

De Queiroz, K. "The ontogenetic method for determining character polarity and its relevance to phylogenetic systematics." *Systematic Zoology* 34 (1985): 280–299.

———. "Ernst Mayr and the modern concept of species." *Proceedings of the United States National Academy of Science* 102 Supplement 1 (2005): 6600–7.

De Robertis, E. M. and Y. Sasai. "A common plan for dorsoventral patterning in Bilateria." *Nature* 380 (1996): 37–40.

Deacon, T.W. "Cortical connections of the inferior arcuate sulcus cortex in the macaque brain." Brain Research 573 (1992): 8–26.

———. *The Symbolic Species: The Co-Evolution of Language and the Brain.* New York: W.W. Norton, 1997.

———. "Evolutionary perspectives on language and brain plasticity." *Journal of Communication Disorders* 33 (2000): 273–90.

———. "Language as Emergent Function: Some Radical Neurological and Evolutionary Implications." In *Religious Narrative, Cognition, and Culture: Image and World in the Mind of Narrative* by A.W. Gertz and J.S. Jensen. London: Equinox Publishing, 2008.

Dean, B. *Fishes, Living and Fossil. An Outline of their Forms and Probable Relationships.* New York, London: Macmillan and Co., 1895.

DeVore, I. "Social behavior and organization of baboons troops." Ph.D. thesis, University of Chicago, 1962.

Dickinson, M. H., C. T. Farley, R. J. Full, et al. "How animals move: an integrative view." *Science* 288 (2000): 100–6.

Dickson, J. A. "Fossil echinoderms as monitor of the Mg/Ca ratio of Phanerozoic oceans." *Science* 298 (2002): 1222–4.

Drew, G.A. "Sexual Attraction in the Squid." *Journal of Morphology* 22 (1911): 327–359.

Eimerl, S., and I. DeVore. *The Primates*. New York: Time-Life Books, 1968.

Elder, H. Y. "Peristaltic mechanisms." In *Aspects of Animal Movement* by H. Y. Elder and E. R. Trueman. Cambridge, U.K., New York: Cambridge University Press, 1980.

Eldredge, N., and S. J. Gould. "Punctuated equilibria: an alternative to phyletic gradualism." In *Models in Paleobiology* by T. J. M. Schopf. San Francisco: Freeman, Cooper & Co., 1972.

Faulkner, R. O., O. Goelet, E. Von Dassow, et al. *The Egyptian Book of the Dead: The Book of Going Forth by Day….* San Francisco: Chronicle Books, 1998.

Filler, A.G. "Anatomical specializations in the hominoid lumbar region." *American Journal of Physical Anthropology* 54 (1981): 218.

———. "Axial Character Seriation in Mammals: An Historical and Morphological Exploration of the Origin, Development, Use and Current Collapse of the Homology Paradigm." Ph.D. thesis, Harvard University, 1986. (Second Printing: Boca Raton, Fla.: Brown Walker Press, 2007.)

———. "Evolution of the sacrum in hominoids." In *Surgical Disorders of the Sacrum* by J. R. Doty and S. S. Rengachary. New York: Thieme Medical Publishers, 1993.

———. "Functional and Evolutionary Perspectives on Chimpanzee Thoracolumbar Musculature." Master's thesis, University of Chicago, 1979.

———. "Homeotic evolution in the Mammalia: Diversification of therian axial seriation and the morphogenetic basis of human origins. *PLoS ONE* (forthcoming), 2007(a).

———. "The emergence and optimization of upright posture among hominiform hominoids and the evolutionary pathophysiology of back pain." (forthcoming), 2007(b).

———. "The first recorded neurosurgical operation—an historical hypothesis: Isis, Osiris, Thoth and the origin of the djed cross spinal symbol." *Neurosurgical Focus* 23 (1): 2007(c).

Filler, A. G., F. A. Howe, C. E. Hayes, et al. "Magnetic resonance neurography." *Lancet* 341 (1993): 659–61.

Filler, A. G., M. Kliot, F. A. Howe, et al. "Application of magnetic resonance neurography in the evaluation of patients with peripheral nerve pathology." *Journal of Neurosurgery* 85 (1996): 299–309.

Filler, A. G., and A. M. Lever. "Effects of cation substitutions on reverse transcriptase and on human immunodeficiency virus production." *AIDS Research and Human Retroviruses* 13 (1997): 291–9.

———. "Molecular evidence for environmental trigger of mass evolutionary acceleration...." Boise, Idaho: American Association for the Advancement of Science, Pacific Division, 2007.

Filler, A. G., K. R. Maravilla, and J. S. Tsuruda. "MR neurography and muscle MR imaging for image diagnosis of disorders affecting the peripheral nerves and musculature." *Neurologic Clinics* 22 (2004): 643–82, vi–vii.

Fisher, R. A. *The Genetical Theory of Natural Selection*. Oxford, U.K.: Clarendon Press, 1930.

Fitch, W. M. "Homology: A personal view on some of the problems." *Trends in Genetics* 16 (2000): 227–31.

Fleagle, J. G. *Primate Adaptation and Evolution*. San Diego: Academic Press, 1999.

Flower, W. H. *An Introduction to the Osteology of the Mammalia: Being the Substance of the Course of Lectures Delivered at the Royal College of Surgeons of England in 1870*. London: Macmillan, 1876.

———. *An Introduction to the Osteology of the Mammalia....* Reprint. London: Macmillan, 1885.

Forterre, P. "The origin of viruses and their possible roles in major evolutionary transitions." *Virus Research* 117 (2006): 5–16.

France. Commission des Sciences et Arts d'Egypte. Napoleon et G. Néret. *Description de l'Egypte*. Köln: Benedikt Taschen, 2002.

France. Commission des Sciences et Arts d'Egypte. *Description de l'Égypte....* Paris: Imprimerie de C.L.F. Panckoucke, 1821.

Fusco, G. "Trunk segment numbers and sequential segmentation in myriapods." *Evolution and Development* 7 (2005): 608–17.

Gambaryan, P. P. *How Mammals Run: Anatomical Adaptations*. New York: Wiley, 1974.

Gebo, D. L. "Climbing, brachiation, and terrestrial quadrupedalism: historical precursors of hominid bipedalism." *American Journal of Physical Anthropology* 101 (1996): 55–92.

———. *Postcranial Adaptation in Nonhuman Primates*. DeKalb, Ill.: Northern Illinois University Press, 1993.

Gebo, D. L., L. MacLatchy, R. Kityo, et al. "A hominoid genus from the Early Miocene of Uganda." *Science* 276 (1997): 401–4.

Geoffroy Saint-Hilaire, E. *Considérations générales sur la vertèbre. Memoires du Museum National d'Histoire Naturelle* 9 (1822): 89–119.

———. *Memoires sur l'organisation des insectes....* Paris: C.L.F. Panckoucke, 1820. (Also published in *Annales Générales des Sciences Physiques* 5 (1820): 96–132.)

———. *Observations sur la concordance des parties de l'hyoide dans le quatre classes animaux vertebras. Nouvelles Annales du Museum d'Histoire Naturelle* 1 (1832): 321–56.

Geoffroy Saint-Hilaire, E., and J. T. E. Hamy. *Lettres écrites d'Égypte....* Paris: Librairie Hachette, 1901.

Geoffroy Saint-Hilaire, I. *Vie, Travaux et Doctrine Scientifique d'Étienne Geoffroy Saint-Hilaire*. Paris: P. Bertrand, 1847.

Ghiselin, M. T. *The Triumph of the Darwinian Method*. Berkeley: University of California Press, 1969.

———. *Metaphysics and the Origin of Species*. Albany: State University of New York Press, 1997.

Gingerich, P. D. "Cetacea." In *The Rise of Placental Mammals: Origins and Relationships of the Major Extant Clades* by K. D. Rose and J. D. Archibald. Baltimore, Md.: Johns Hopkins University Press, 2005. 234–52.

Goethe, J. W. v. *Dem Menschen wie den Tieren ist ein Zwischenknochen der odem Kinnlade zuzuschreiben (An Intermaxillary Bone is Present in the Upper Jaw of Man as well as in Animals). Sämtliche Werke* 2 (1784): 530–45.

———. *Italian Journey, 1786-1788*. Translated by W. H. Auden. Harmondsworth: Penguin Books, 1970.

———. *Zur Naturwissenschaft überhaupt, besonders zur Morphologie*. Stuttgard: Tübingen, 1817.

Goethe, J. W. v., J. P. Eckermann, and M. Fuller. *Conversations with Goethe in the Last Years of His Life*. Boston: J. Munroe, 1852.

Gommery, D. "Evolution of the vertebral column in Miocene hominoids and Plio-Pleistocene hominids." In *Human Origins and Environmental Backgrounds*. H. Ishida, R. H. Tuttle, M. Pickford, N. Ogihara and M. Nakatsukasa. New York: Springer, 2006. 31–44.

———. "From head to the pelvis or the evolution of the vertebral column in Miocene hominoids and Plio-Pleistocene hominids." *Abstracts of the International Symposium on Human Origins and Environmental Backgrounds*. Kyoto, Japan: 2003.

Gordon, A. H. and C. W. Schwabe. *The Quick and the Dead: Biomedical Theory in Ancient Egypt*. Boston: Brill Academic Publishers, 2004.

Gould, S. J. "The exaptive excellence of spandrels as a term and prototype." *Proceedings of the United States National Academy of Science* 94 (1997): 10750–5.

———. *The Structure of Evolutionary Theory*. Cambridge, Mass.: Belknap Press of Harvard University Press, 2002.

Gould, S. J., and R. C. Lewontin. "The spandrels of San Marco and the Panglossian paradigm: a critique of the adaptationist programme." *Proceedings of the Royal Society of London, Series B—Biological Science* 205 (1979): 581–98.

Griffiths, A. J. F. *Introduction to Genetic Analysis*. New York: W.H. Freeman, 2005.

Gruneberg, H. "Genetical studies on the skeleton of the mouse. II. *Undulated* and its 'modifiers.'" *Journal of Genetics* 50 (1950): 142–73.

———. "Genetical studies on the skeleton of the mouse. XII. The development of *undulated*." *Journal of Genetics* 52 (1954): 441–55.

Guy, F., D. E. Lieberman, D. Pilbeam, et al. "Morphological affinities of the *Sahelanthropus tchadensis* (Late Miocene hominid from Chad) cranium." *Proceedings of the United States National Academy of Science* 102 (2005): 18836–41.

Haeusler, M., S. A. Martelli, and T. Boeni. "Vertebrae numbers of the early hominid lumbar spine." *Journal of Human Evolution* 43 (2002): 621–43.

Harcourt-Smith, W. E. and L. C. Aiello. "Fossils, feet and the evolution of human bipedal locomotion." *Journal of Anatomy* 204 (2004): 403–16.

Harrison, G. P., M. S. Mayo, E. Hunter, et al. "Pausing of reverse transcriptase on retroviral RNA templates is influenced by secondary structures both 5' and 3' of the catalytic site." *Nucleic Acids Research* 26 (1998): 3433–42.

Hart, C. P., A. Awgulewitsch, A. Fainsod, et al. "Homeo box gene complex on mouse chromosome 11: molecular cloning, expression in embryogenesis, and homology to a human homeobox locus." *Cell* 43 (1985): 9–18.

Haycock, D. B. "Ancient Egypt in 17th and 18th Century England." In *The Wisdom of Egypt: Changing Visions Through the Ages* by P. J. Ucko and T. Champion. London: U.C.L. Press, Cavendish Publishing Ltd., 2003. 133–60.

Heilmann, G. *The Origin of Birds*. New York: D. Appleton and Company, 1927.

Hennig, W. *Phylogenetic Systematics*. Urbana, Ill., University of Illinois Press, 1966.

Hickman, C. *Integrated Principals of Zoology*. Saint Louis: Mosby, 1955.

Hickman, C. P. *Biology of the Invertebrates*. Saint Louis: Mosby, 1973.

Hobbes, T., A. Crooke, et al. *Leviathan; or, The matter, forme, & power of a common-wealth ecclesiasticall and civill*. London: Printed for Andrew Crooke at the Green Dragon in St. Paul's Church-yard, 1651.

Hoffstetter, R., and J.P. Gasc. "Vertebrae and ribs of modern reptiles." In *Biology of the Reptilia—Morphology* by A. C. Gans, A. D. A. Bellairs, and T. S. Parsons. London: Academic Press, 1969. 201–310.

Huang, J., Y. Xu, and J. P. Gogarten. "The presence of a haloarchaeal type tyrosyl-tRNA synthetase marks the opisthokonts as monophyletic." *Molecular Biology and Evolution* 22 (2005): 2142–6.

Hughes, N. C., and D. K. Jacobs. "The end of everything: metazoan terminal addition." *Evolution and Development* 7 (2005): 497.

Ikuta, T., and H. Saiga. "Organization of *Hox* genes in ascidians: present, past, and future." *Developmental Dynamics* 233 (2005): 382–9.

Irschick, D. J., and B. C. Jayne. "Comparative three-dimensional kinematics of the hindlimb for high-speed bipedal and quadrupedal locomotion of lizards." *Journal of Experimental Biology* 202 (1999): 1047–65.

Ishida, H., Y. Kunimatsu, T. Takano, et al. "*Nacholapithecus* skeleton from the Middle Miocene of Kenya." *Journal of Human Evolution* 46 (2004): 69–103.

Jacobs, D. K., N. C. Hughes, S. T. Fitz-Gibbon, et al. "Terminal addition, the Cambrian radiation and the Phanerozoic evolution of bilaterian form." *Evolution and Development* 7 (2005): 498–514.

Janensch, W. "Die Wirbelsäule von *Brachiosaurus brancai*." *Palaeontographica Supplement* VII, Erste Reile, II (1950): 27–93.

Jenkins, F. A. "The Chanares (Argentina) Triassic reptile fauna. VII. The postcranial skeleton of the traversodontid *Masetognathus pascuali* (Therapsida, Cynodontia)." *Breviora* 352 (1970): 1–28.

———. "The Postcranial Skeleton of African Cynodonts; Problems in the Early Evolution of the Mammalian Postcranial Skeleton." New Haven: Peabody Museum of Natural History, Yale University, 1971.

Jenkins, F. A., Jr. "Chimpanzee bipedalism: cineradiographic analysis and implications for the evolution of gait." *Science* 178 (1972): 877–9.

Jenkins, F. A., and J. G. Fleagle. "Knuckle-walking and the functional anatomy of the wrists in living apes." In *Primate Functional Morphology and Evolution* by R. Tuttle. Chicago: Mouton/Aldine, 1975. 213–27.

Johanson, D. C., and M. Taieb. "Plio-Pleistocene hominid discoveries in Hadar, Ethiopia." *Nature* 260 (1976): 293–7.

Kardong, K. V. *Vertebrates: Comparative Anatomy, Function, Evolution, 4th Edition*. New York: McGraw-Hill Higher Education, 2005.

Kehrer-Sawatzki, H., J. M. Szamalek, S. Tanzer, et al. "Molecular characterization of the pericentric inversion of chimpanzee chromosome 11 homologous to human chromosome 9." *Genomics* 85 (2005): 542–50.

Keith, A. "The extent to which the posterior segments of the body have been transmuted and suppressed in the evolution of man and allied primates." *Journal of Anatomy and Physiology* 37 (1902): 18–40.

———. "Man's posture: its evolution and disorders." In the Hunterian Lectures I-VI. *British Medical Journal* (1923): 451–54; 499–502; 545–48; 587–90; 624–26; 669–72.

Kohler, M., and S. Moya-Sola. "Ape-like or hominid-like? The positional behavior of *Oreopithecus bambolii* reconsidered." *Proceedings of the United States National Academy of Science* 94 (1997): 11747–50.

Kokubu, C., B. Wilm, T. Kokubu, et al. "*Undulated* short-tail deletion mutation in the mouse ablates *Pax1* and leads to ectopic activation of neighboring *Nkx2-2* in domains that normally express *Pax1*." *Genetics* 165 (2003): 299–307.

Koonin, E. V., T. G. Senkevich and V. V. Dolja. "The ancient Virus World and evolution of cells." *Biology Direct* 1 (2006): 29.

Kuratani, S. "Craniofacial development and the evolution of the vertebrates: the old problems on a new background." *Zoological Science* 22 (2005a): 1–19.

———. "Developmental studies of the lamprey and hierarchical evolutionary steps towards the acquisition of the jaw." *Journal of Anatomy* 207 (2005b): 489–99.

Lang, J. K., and H. Kolenda. "First appearance and sense of the term 'spinal column' in ancient Egypt." *Journal of Neurosurgery* 97 Spine 1 (2002): 152–55.

Le Guyader, H. *Geoffroy Saint-Hilaire: A Visionary Naturalist*. Chicago, Ill.: University of Chicago Press, 2004.

Lee, P. N., P. Callaerts, H. G. De Couet, et al. "Cephalopod *Hox* genes and the origin of morphological novelties." *Nature* 424 (2003): 1061–5.

Lewis, O. J. "The contrasting morphology found in the wrist joints of semibrachiating monkeys and brachiating apes." *Folia Primatologica* (Basel) 16 (1971): 248–56.

———. "Derived morphology of the wrist articulations and theories of hominoid evolution: Part II. The midcarpal joints of higher primates." *Journal of Anatomy* 142 (1985): 151–72.

———. *Functional Morphology of the Evolving Hand and Foot*. Oxford, U.K.; New York: Clarendon Press, Oxford University Press, 1989. "The hominoid os capitulum, with special reference to the fossil bones from Sterkfontein and Olduvai Gorge." *Journal of Human Evolution* 2 (1973): 1–11.

———. "Joint remodelling and the evolution of the human hand." *Journal of Anatomy* 123 Pt. 1 (1977): 157–201.

Lichtneckert, R., and H. Reichert. "Insights into the urbilaterian brain: conserved genetic patterning mechanisms in insect and vertebrate brain development." *Heredity* 94 (2005): 465–77.

Liem, K., W. Bemis, W. F. Walker, et al. *Functional Anatomy of the Vertebrates: An Evolutionary Perspective*, 3rd Ed. Boston: Brooks Cole, 2001.

Liu, P. Z., and T. C. Kaufman. "Short and long germ segmentation: unanswered questions in the evolution of a developmental mode." *Evolution and Development* 7 (2005): 629–46.

Lovejoy, C. O. "The natural history of human gait and posture. Part 1. Spine and pelvis." *Gait & Posture* 21 (2005a): 95–112.

———. "The natural history of human gait and posture. Part 2. Hip and thigh." *Gait & Posture* 21 (2005b): 113-24.

———. "The natural history of human gait and posture Part 3. The knee." *Gait & Posture* 25 (2007): 325–41.

Lovejoy, C. O., K. G. Heiple, and R. S. Meindl. "Palaeoanthropology: Did our ancestors knuckle-walk?" *Nature* 410 (2001): 325–6.

Lurker, M. *The Gods and Symbols of Ancient Egypt: An Illustrated Dictionary*. New York: Thames and Hudson, 1980.

MacLatchy, L. "The oldest ape." *Evolutionary Anthropology* 13 (2004): 90–103.

Mansouri, A., A. K. Voss, T. Thomas, et al. "*Uncx4.1* is required for the formation of the pedicles and proximal ribs and acts upstream of *Pax9*." *Development* 127 (2000): 2251–8.

Mayr, E. *What Evolution Is*. New York: Basic Books, 2001.

Mayr, E., and W. B. Provine. *The Evolutionary Synthesis: Perspectives on the Unification of Biology.* Cambridge, Mass.: Harvard University Press, 1998.

McFadden, G. I., and G. G. Van Dooren. "Evolution: red algal genome affirms a common origin of all plastids." *Current Biology* 14 (2004): R514–6.

McGinnis, W., R. L. Garber, J. Wirz, et al. "A homologous protein-coding sequence in *Drosophila* homeotic genes and its conservation in other metazoans." *Cell* 37 (1984a): 403–8.

McGinnis, W., C. P. Hart, W. J. Gehring, et al. "Molecular cloning and chromosome mapping of a mouse DNA sequence homologous to homeotic genes of *Drosophila.*" *Cell* 38 (1984b): 675–80.

McGinnis, W., M. Levine, E. Hefen, et al. "A conserved DNA sequence in homeotic genes of the *Drosophila Antennapedia* and *bithorax* complexes." *Nature* 308 (1984c): 428–33.

McLoughlin, J. C. *Archosauria, A New Look at the Old Dinosaur.* New York: Viking Press, 1979.

Meier, S. "Development of the chick embryo mesoblast. Formation of the embryonic axis and establishment of the metameric pattern." *Developmental Biology* 73 (1979): 25–45.

Meier, S. and P. P. Tam. "Metameric pattern development in the embryonic axis of the mouse. I. Differentiation of the cranial segments." *Differentiation* 21 (1982): 95–108.

Mikkelson, T. S., et al. "Initial sequence of the chimpanzee genome and comparison with the human genome." *Nature* 437 (2005): 69–87.

Minelli, A., and G. Fusco. "Conserved versus innovative features in animal body organization." *Journal of Experimental Zoology, Part B—Molecular and Developmental Evolution* 304 (2005): 520–5.

———. "Evo-devo perspectives on segmentation: model organisms, and beyond." *Trends in Ecology and Evolution* 19 (2004): 423–9.

Moya-Sola, S., M. Kohler, D. M. Alba, et al. "*Pierolapithecus catalaunicus*, a new Middle Miocene great ape from Spain." *Science* 306 (2004): 1339–44.

Moya-Sola, S., M. Kohler, and L. Rook. "Evidence of hominid-like precision grip capability in the hand of the Miocene ape *Oreopithecus. Proceedings of the United States National Academy of Science* 96 (1999): 313–7.

Müller, G. "Homology: The evolution of morphological organization." In *Origination of Organismal Form: Beyond the Gene in Developmental and Evolutionary Biology* by G. Müller and S. Newman. Cambridge, Mass.: MIT Press, 2003. 51–70.

Nakatsukasa, M., H. Tsujikawa, D. Shimizu, et al. "Definitive evidence for tail loss in *Nacholapithecus*, an East African Miocene hominoid." *Journal of Human Evolution* 45 (2003): 179–86.

Napier, J. R., and P. H. Napier. *The Natural History of the Primates.* Cambridge, Mass.: MIT Press, 1985.

Noden, D. M., and P. A. Trainor. "Relations and interactions between cranial mesoderm and neural crest populations." *Journal of Anatomy* 207 (2005): 575–601.

Ohman, J. C., C. O. Lovejoy, T. D. White, et al. "Questions about Orrorin femur." *Science* 307 (2005): 845b.

Oken, L., and A. Tulk. *Elements of Physiophilosophy*. London: Printed for the Ray Society, 1847.

Owen, R. *On the Archetype and Homologies of the Vertebrate Skeleton*. London: J. Van Voorst, 1848.

———. *On the classification and geographical distribution of the Mammalia....* London: J. W. Parker, 1859.

———. *Instances of the power of God as manifested in his animal creation*. London: Simpkin & Marshall, 1864.

———. "Lectures on the comparative anatomy and physiology of the vertebrate animals, delivered at the Royal college of surgeons of England, in 1844 and 1846." London: Printed for Longman Brown Green and Longmans, 1846.

———. "Osteological contributions to the natural history of the chimpanzees (*Troglodytes*) and orangs (*Pithecus*). No. V. Comparison of the lower jaw and vertebral column of the *Troglodytes gorilla*, *Troglodytes niger*, *Pithecus satyrus*, and different varieties of the human race." Transactions of the Zoological Society of London, 4 (1857): 89–115. Printed in 1862.

———. "Report on British fossil reptiles. Part II." Report of the Eleventh Meeting of the British Association for the Advancement of Science. Plymouth, U.K., 1842.

Padian, K. "A missing Hunterian Lecture on vertebrae by Richard Owen, 1837." *Journal of the History of Biology* 28 (1995): 333–68.

Pearson, K. *The Life, Letters and Labours of Farncis Galton*. Cambridge, U.K.: Cambridge University Press, 1914.

Penny, D., and M. J. Phillips. "The rise of birds and mammals: are microevolutionary processes sufficient for macroevolution?" *Trends in Ecology and Evolution* 19 (2004): 516–22.

Perry, J., S. Nouri, P. La, et al. "Molecular distinction between true centric fission and pericentric duplication-fission." *Human Genetics* 116 (2005): 300–10.

Peters, H., B. Wilm, N. Sakai, et al. "*Pax1* and *Pax9* synergistically regulate vertebral column development." *Development* 126 (1999): 5399–408.

Pilbeam, D. "The anthropoid postcranial axial skeleton: comments on development, variation, and evolution." *Journal of Experimental Zoology, Part B—Molecular and Developmental Evolution* 302 (2004): 241–67.

———. "Tertiary Pongidae of East Africa: evolutionary relationships and taxonomy." New Haven: Peabody Museum of Natural History, Yale University, 1969.

Pinch, G. *Egyptian Mythology: A Guide to the Gods, Goddesses, and Traditions of Ancient Egypt*. Oxford, U.K.; New York: Oxford University Press, 2004.

Plato, *The Republic*, ed. A. D. Bloom. New York: Basic Books, 1991.

Plutarch, *Moralia, Volume V, Isis and Osiris*, trans. Frank Cole Babbitt. Cambridge, Mass.: Loeb Classical Library, Harvard University Press, 1936.

Prangishvili, D., P. Forterre and R. A. Garrett. "Viruses of the Archaea: a unifying view." *Nature Reviews—Microbiology* 4 (2006): 837–48.

Preuschoft, H. "Mechanisms for the acquisition of habitual bipedality: are there biomechanical reasons for the acquisition of upright bipedal posture?" *Journal of Anatomy* 204 (2004): 363–84.

Raaum, R. L., K. N. Sterner, C. M. Noviello, et al. "Catarrhine primate divergence dates estimated from complete mitochondrial genomes: concordance with fossil and nuclear DNA evidence." *Journal of Human Evolution* 48 (2005): 237–57.

Randall, D. J., W. W. Burggren, A. P. Farrell, et al. *The Evolution of Air Breathing in Vertebrates*. Cambridge, U.K.; New York: Cambridge University Press, 1981.

Raven, P.H., G.B. Johnson, J.B. Losos et al. *Biology*. Boston: McGraw-Hill, 2005.

Reichert, H. and A. Simeone. "Developmental genetic evidence for a monophyletic origin of the bilaterian brain." *Philosophical Transactions of the Royal Society of London, Series B—Biological Science* 356 (2001): 1533–44.

Richards, R. J. *The Romantic Conception of Life: Science and Philosophy in the Age of Goethe*. Chicago: University of Chicago Press, 2002.

Richmond, B. G. "Functional morphology of the midcarpal joint in knuckle-walkers and terrestrial quadrupeds." In *Human Origins and Environmental Backgrounds* by H. Ishida, R. H. Tuttle, M. Pickford, N. Ogihara, and M. Nakatsukasa. New York: Springer, 2006. 105–122.

Richmond, B. G., D. R. Begun, and D. S. Strait. "Origin of human bipedalism: The knuckle-walking hypothesis revisited." *American Journal of Physical Anthropology* Supplement 33 (2001): 70–105.

Richmond, B. G., and D. S. Strait. "Evidence that humans evolved from a knuckle-walking ancestor." *Nature* 404 (2000): 382–5.

Ritter, D.A., P.N. Nassar, M. Fife et al. "Epaxial muscle function in trotting dogs." *Journal of Experimental Biology* 204 (2001): 2053–64.

Robertson, J. T. *Early Hominid Posture and Locomotion*. Chicago: University of Chicago Press, 1972.

Robinson, N. P., and S. D. Bell. "Origins of DNA replication in the three domains of life." *Federation of European Biochemical Societies Journal* 272 (2005): 3757-66.

Romer, A. S. *Osteology of the Reptiles*. Chicago, University of Chicago Press, 1956.

Romer, A. S., and A. D. Lewis. "The Chanares (Argentina) Triasic reptile fauna. XIX. Postcranial materials of the Cynodonts *Probelesdon* and *Probainognathus*." *Breviora* 407 (1973): 1–26.

Rook, L., L. Bondioli, M. Kohler, et al. "*Oreopithecus* was a bipedal ape after all: evidence from the iliac cancellous architecture." *Proceedings of the United States National Academy of Science* 96 (1999): 8795–9.

Rutherford, E., and F. Soddy. "The cause and nature of radioactivity." *Philosophical Magazine* 4 (1902): 370–396.

Ruvolo, M. "Comparative primate genomics: the year of the chimpanzee." *Current Opinion in Genetics and Development* 14 (2004): 650–6.

———. "Molecular evolutionary processes and conflicting gene trees: the hominoid case." *American Journal of Physical Anthropology* 94 (1994): 89–113.

———. "Molecular phylogeny of the hominoids: inferences from multiple independent DNA sequence data sets." *Molecular Biology and Evolution* 14 (1997): 248–65.

———. "A new approach to studying modern human origins: hypothesis testing with coalescence time distributions." *Molecular Phylogenetics and Evolution* 5 (1996): 202–19.

Ruvolo, M., T. R. Disotell, M. W. Allard, et al. "Resolution of the African hominoid trichotomy by use of a mitochondrial gene sequence." *Proceedings of the United States National Academy of Science* 88 (1991): 1570–4.

Sagai, T., H. Masuya, M. Tamura et al. "Phylogentic conservation of a limb-specific, cis-acting regulator Sonic hedgehog (Shh)." *Mammalian Genome* 15 (2004): 23–34.

Sanchez-Garcia, I., H. Osada, A. Forster, et al. "The cysteine-rich LIM domains inhibit DNA binding by the associated homeodomain in Isl-1." *European Molecular Biology Organization Journal* 12 (1993): 4243–50.

Sanders, W. J. and B. E. Bodenbender. "Morphometric analysis of lumbar vertebra UMP 67-28: Implications for the Miocene Moroto hominoid." *Journal of Human Evolution* 26 (1994): 203–237.

Schache, A. G., P. Blanch, D. Rath, et al. "Three-dimensional angular kinematics of the lumbar spine and pelvis during running." *Human Movement Science* 21 (2002): 273–93.

Schilling, N., and R. Hackert. "Sagittal spine movements of small therian mammals during asymmetrical gaits." *Journal of Experimental Biology* 209 (2006): 3925–39.

Schlosser, G., and G. P. Wagner. *Modularity in Development and Evolution*. Chicago: University of Chicago Press, 2004.

Schmidt-Nielsen, K. *Animal Physiology: Adaptation and Environment*. Cambridge, U.K.; New York: Cambridge University Press, 1979.

Schultz, A. H. "Die Körporproportionen der erwachsenen catarrhinen Primaten, mit spezieller Berüchsichtigung der Menschenaffen." *Anthropologischer Anzeiger* 10 (1933): 154–85.

———. *The Life of Primates*. New York: Universe Books, 1969.

Shapiro, L. "Functional morphology of the vertebral column in primates." In *Postcranial Adaptation in Nonhuman Primates* by D. L. Gebo. Dekalb, Ill.: Northern Illinois University Press, 1993. 121–49.

Silcox, M. T., J. I. Bloch, E. J. Sargis, et al. "Euarchonta (Dermoptera, Scandentia, Primates)." In *The Rise of the Placental Mammals: Origins and Relationships of the Major Extant Clades* by K. D. Rose and J. D. Archibald. Baltimore: Johns Hopkins University Press, 2005. 127–44.

Simons, E. L. *Primate Evolution: An Introduction to Man's Place in Nature*. New York: Macmillan, 1972.

Simpson, G. G. *The Major Features of Evolution*. New York: Columbia University Press, 1953.

———. *Tempo and Mode in Evolution*. New York: Columbia University Press, 1944.

Sleigh, M. A. and D. I. Barlow. "Metachronism and control of locomotion in animals with many propulsive structures." In *Aspects of Animal Movement*, by H. Y. Elder and E. R. Trueman. Cambridge, U.K.; New York: Cambridge University Press, 1980. 49–70.

Smith, T., K. D. Rose, and P. D. Gingerich. "Rapid Asia-Europe-North America geographic dispersal of earliest Eocene primate *Teilhardina* during the Paleocene-Eocene Thermal Maximum." *Proceedings of the United States National Academy of Science* 103 (2006): 11223–7.

Stanley, S. M. *Macroevolution, Pattern and Process*. Baltimore: Johns Hopkins University Press, 1998.

Stanley, S. M., J. B. Ries and L. A. Hardie. "Low-magnesium calcite produced by coralline algae in seawater of Late Cretaceous composition." *Proceedings of the United States National Academy of Science* 99 (2002): 15323–6.

Steenkamp, E. T., J. Wright, and S. L. Baldauf. "The protistan origins of animals and fungi." *Molecular Biology and Evolution* 23 (2006): 93–106.

Steiner, R., J. W. V. Goethe, and J. Barnes. *Nature's Open Secret: Introductions to Goethe's Scientific Writings*. Great Barrington, Mass.: Anthroposophic Press, 2000.

Stern, J. T. "Before bipedality." *Yearbook of Physical Anthropology* 19 (1975): 59–68.

Stollewerk, A., M. Schoppmeier, and W. G. Damen. "Involvement of *Notch* and *Delta* genes in spider segmentation." *Nature* 423 (2003): 863–5.

Stoltzfus, A. "Mutationism and the dual causation of evolutionary change." *Evolution and Development* 8 (2006): 304–17.

Struhl, G. "A homoeotic mutation transforming leg to antenna in *Drosophila*." *Nature* 292 (1981): 635–8.

Tam, P. P., S. Meier, and A. G. Jacobson. "Differentiation of the metameric pattern in the embryonic axis of the mouse. II. Somitomeric organization of the presomitic mesoderm." *Differentiation* 21 (1982): 109–22.

Tuttle, R. H. "Are human beings apes, or are apes people too?" In *Human Origins and Environmental Backgrounds*, by H. Ishida, R. H. Tuttle, M. Pickford, N. Ogihara, and M. Nakatsukasa. New York: Springer, 2006a. 248–58.

———. "Darwin's apes, dental apes, and the descent of Man: Normal science in evolutionary anthropology." *Current Anthropology* 15 (1974): 389–98.

———. "Knuckle-walking and the evolution of hominoid hands." *American Journal of Physical Anthropology* 26 (1967): 171–206.

———. "Knuckle-walking and knuckle-walkers: A commentary on some recent perspectives on hominoid evolution." In *Primate Functional Morphology and Evolution*. Chicago: Mouton/Aldine, 1975. 203–9.

———. "Knuckle-walking and the problem of human origins." *Science* 166 (1969): 953–61.

———. "Neontological perspectives on East African Middle and Late Miocene Anthropoidea." In *Human Origins and Environmental Backgrounds*, by H. Ishida, R. H. Tuttle, M. Pickford, N. Ogihara, and M. Nakatsukasa. New York: Springer, 2006b. 209–23.

———. "Phylogenies, fossils, and feelings." In *Great Apes and Humans: The Ethics of Coexistence*, by B. B. Beck, T. S. Stoinski, A. Arlukeet, et al. Washington, D.C.: Smithsonian Institution Press, 2001. 178–90.

———. "Seven decades of East African Miocene anthropoid studies." In *Human Origins and Environmental Backgrounds*, by H. Ishida, R. H. Tuttle, M. Pickford, N. Ogihara, and M. Nakatsukasa. New York: Springer, 2006c. 15–30.

Tuttle, R. H., M. J. Velte, and J. V. Basmajian. "Electromyography of brachial muscles in *Pan troglodytes* and *Pongo pygmaeus*." *American Journal of Physical Anthropology* 61 (1983): 75–83.

Vereecke, E.E., K. D'Aout, L. Van Elsacker et al. "Functional analysis of the gibbon foot during terrestrial bipedal walking...." *American Journal of Physical Anthropology* 128 (2005): 659–69.

Vereecke, E. E., K. D'Aout, and P. Aerts. "Locomotor versatility in the white-handed gibbon (*Hylobates lar*): a spatiotemporal analysis of the bipedal, tripedal, and quadrupedal gaits." *Journal of Human Evolution* 50 (2006): 552–67.

Volpe, M.V., R.J. Vosatka, and H.C. Nelson. "*Hoxb-5* control of early airway formation during branching morphogenesis in the developing mouse lung." *Biochimica et Biophysica Acta (BBA) General Subjects* 1475 (2000): 337–45.

Wachtershauser, G. "From volcanic origins of chemoautotrophic life to Bacteria, Archaea and Eukarya." *Philosophical Transactions of the Royal Society of London, Series B—Biological Science* 361 (2006): 1787-806; discussion 1806–8.

Walker, A., and M. D. Rose. "Fossil hominoid vertebra from the Miocene of Uganda." *Nature* 217 (1968): 980–1.

Wallin, J., J. Wilting, H. Koseki, et al. "The role of *Pax-1* in axial skeleton development." *Development* 120 (1994): 1109–21.

Wang, W. J., R. H. Crompton, Y. Li, et al. "Energy transformation during erect and 'bent-hip, bent-knee' walking by humans with implications for the evolution of bipedalism." *Journal of Human Evolution* 44 (2003): 563-79.

Ward, C. V., A. Walker, M. F. Teaford, et al. (1993). "Partial skeleton of *Proconsul nyanzae* from Mfangano Island, Kenya." *American Journal of Physical Anthropology* 90 (1993): 77–111.

Washburn, S. L. "The analysis of primate evolution with particular reference to the origin of man." *Cold Spring Harbor Symposia on Quantitative Biology* 15 (1950): 67–78.

———. "The evolution of man." *Scientific American* 239 (1978): 194–8, 201–2, 204 passim.

———. "Human evolution." *Perspectives in Biology and Medicine* 25 (1982): 583–602.

Watson, J. D., and F. H. Crick. "Molecular structure of nucleic acids; a structure for deoxyribose nucleic acid." *Nature* 171 (1953): 737–8.

Wettstein, O. "Crocodilia." In *Handbuch der Zoologie* by W. Kukenthal and T. Krumbach. Berlin: W. de Gruyter & Co., 1937. VII: 236–320.

Wheeler, Q., and R. Meier. *Species Concepts and Phylogenetic Theory: A Debate.* New York: Columbia University Press, 2000.

White, T. D., G. WoldeGabriel, B. Asfaw, et al. "Asa Issie, Aramis and the origin of *Australopithecus.*" *Nature* 440 (2006): 883–9.

Williston, S. W., and W. K. Gregory. *The Osteology of the Reptiles.* Cambridge, Mass.: Harvard University Press, 1925.

Woese, C. R. "On the evolution of cells." *Proceedings of the United States National Academy of Science* 99 (2002): 8742–7.

———. "A new biology for a new century." *Microbiology and Molecular Biology Reviews* 68 (2004): 173–86.

Woese, C. R., and G. E. Fox. "Phylogenetic structure of the prokaryotic domain: the primary kingdoms." *Proceedings of the United States National Academy of Science* 74 (1977): 5088–90.

Wolpoff, M.H., B. Senut, M. Pickford et al. "Palaeoanthropology. Sahelanthropus or 'Sahelpithicus'?" *Nature* 419 (2002): 581–2.

Wood, B. "Hominid revelations from Chad." *Nature* 418 (2002): 133–5.

Wyss, A. R., M. J. Novacek, and M. C. McKenna. "Amino acid sequence versus morphological data and the interordinal relationships of mammals." *Molecular Biology and Evolution* 4 (1987):99–116.

Yamane, A. "Embryonic and postnatal development of masticatory and tongue muscles." *Cell and Tissue Research* 322 (2005): 183–9.

Young, N. M., and L. MacLatchy. "The phylogenetic position of *Morotopithecus.*" *Journal of Human Evolution* 46 (2004): 163–84.

Zollikofer, C. P., M. S. Ponce de Leon, D. E. Lieberman, et al. "Virtual cranial reconstruction of *Sahelanthropus tchadensis.*" *Nature* 434 (2005): 755–9.

Index

About the Author

Dr. Aaron G. Filler, M.D., Ph.D., studied evolutionary theory under some of the leading biologists of our time, such as Ernst Mayr, Stephen J. Gould, David Pilbeam, and Irven DeVore. Some of his pioneering research at Harvard University dealt with the way in which evolution reveals itself through the variety of shapes and forms of animals and human ancestors. This research was recently published in *Axial Character Seriation in Mammals* (BrownWalker Press), as well as in a variety of academic journals. He is the author of *Do You Really Need Back Surgery?* (Oxford University Press), and has recently been appointed as a co-editor of the *Youman's Textbook of Neurosurgery, 6th edition*. A neurosurgeon at the Institute for Spinal Disorders at Cedars Sinai Medical Center, and past Associate Director of the Comprehensive Spine Center at UCLA, Dr. Filler has been a leading innovator in evolutionary biology, medical imaging, surgery, and neuroscience. He is also the author of numerous scientific articles and patents, as well as a recipient of a variety of research grants. Dr. Filler is one of the very few Americans to be honored by nomination to the Royal College of Surgeons of England—an institution that has been the birthplace of many key elements of our understanding of evolution.

Public interest in his medical imaging and pain treatment research has led to coverage of his work in the *New York Times*, the *Los Angeles Times*, the *Economist*, and the *London Times*. He has appeared on CNN and ABC World News with Peter Jennings, and has been interviewed on numerous local television and national radio channels. He is a frequent invited speaker at major national meetings of orthopedic surgeons, neurosurgeons, neurologists, and other medical groups. He resides in Santa Monica, California.

About the Translators

Elizabeth Jackson is a writer and linguist formerly based at the Getty Museum in Los Angeles who has studied language and literature at the Goethe Institute in Frankfurt and the Herder Institute in Leipzig. Daniel Schaeffer is a native German speaking physician with a background in philosophy who is also fluent in English.